HIGHER BIOLOGY

Bill Dickson and
Graham Moffat

A MEMBER OF THE HODDER HEADLINE GROUP

The Publishers would like to thank the following for permission to reproduce copyright material:

Every effort has been made to trace all copyright holders, but if any have been inadvertently overlooked the Publishers will be pleased to make the necessary arrangements at the first opportunity.

Orders: please contact Bookpoint Ltd, 130 Milton Park, Abingdon, Oxon OX14 4SB. Telephone: (44) 01235 827720. Fax: (44) 01235 400454. Lines are open from 9.00 – 5.00, Monday to Saturday, with a 24-hour message answering service. Visit our website at www.hoddereducation.co.uk. Hodder Gibson can be contacted direct on: Tel: 0141 848 1609; Fax: 0141 889 6315; email: hoddergibson@hodder.co.uk

First published in 2006 by
Hodder Gibson, a member of the Hodder Headline Group
2a Christie Street
Paisley PA1 1NB

Impression number 10 9 8 7 6 5 4 3 2 1
Year 2010 2009 2008 2007 2006

Cover photo by David Aubrey/Science Photo Library
Typeset in 9.5/12.5pt Frutiger Light by Phoenix Photosetting, Chatham, Kent
Printed and bound in Great Britain by Martins The Printers, Berwick-upon-Tweed

A catalogue record for this title is available from the British Library

ISBN-10: 0-340-90608-1
ISBN-13: 978-0-340-90608-8

CONTENTS

INTRODUCTION

General Introduction

Welcome to *How To Pass Higher Biology!*

The fact you have opened this book and are reading it shows that you are keen to pass your Higher Biology Course this year. This is excellent because passing, and passing well, needs that type of attitude. It also shows you are getting down to the revision that is **essential** to pass and get the best grade possible.

The idea behind this book is to help you to pass, and if you are already on track to pass, it can help you improve your grade. It can help boost a C pass into a B pass or a B into an A. It can't do the work for you, but it can guide you in how best to use your limited time.

In producing this book we have assumed that you have followed an SQA Higher Level Biology Course at school or college this year and that you have probably, but not necessarily, done either Standard Grade or Intermediate 2 Biology before.

We recommend that you download and print a copy of the **Higher Level Biology Arrangements** from the SQA internet site at www.sqa.org.uk. Choose **Biology** from the subject list, then, click on **Higher Level Arrangements** and print pages 6–35. **Note that in your exam only the Content and Notes columns of the Arrangements can be examined.** You should also get a copy of a booklet of **Past Papers.**

We suggest that you **use this book throughout** your Course. Use the book at the end of each Topic Area covered in class, at the end of each Unit in preparation for your Unit Test (NAB), before your preliminary examination and, finally, to revise the whole Course in the lead up to your examination.

We have tried to keep the language simple but we **have** used the language of the published Arrangements for Higher Biology for the simple reason that this is the language used in the setting of the examination papers – the examiners **must** use this language.

Although we have covered the entire Higher Biology Course within these materials, we have tried to emphasise those areas that cause most difficulty for candidates.

About This Book

The **Knowledge and Understanding Section** comprises three chapters which cover the three Units of Higher Biology. Each chapter is divided into **Topic Areas**. Each topic area has four features.

Key Ideas

These list and expand the content statements from the Arrangements using the words and phrases needed to answer examination questions. **After having worked on the topic**, these should be easy to understand. You might want to use the boxes to show progress. We suggest marking like this ⊟ if you are having difficulty, like this ⊞ if you have done further work and are more comfortable and finally, like this ⊠ once you are confident and have **learned** a particular idea. Alternatively, you could traffic light them – a red dot for 'not understood', an orange dot for 'more work needed' and a green dot for 'fully understood'.

Topic Notes

These paragraphs give a summary of the knowledge required in the topic area and should be read carefully. You might like to use a highlighter pen to emphasise certain parts or you might add extra information from your class notes. The labels on diagrams should be read carefully.

Top Tip

Where we offer a tip to help learning it is shown in a special box like this. The tips can be very general or can be specific to the content of the topic area. The specific content tips are only **suggestions** – don't feel you need to use them all!

Exam Questions

These questions have been carefully produced to show typical questions that you will be asked in Section B and Section C of your examination paper. We have marked each question as either C-type which most candidates who pass the exam will get right, B-type which are more difficult or A-type which are answered well by candidates who go on to achieve A grades.

The questions cover a sample of the knowledge from each topic area and include structured examples and an example of each type of extended response question. You should try these questions on separate paper then mark your answers. SQA standard answers are provided in the Exam Questions: Answers and Commentary chapter.

The **Problem Solving** chapter covers the seven main problem solving and practical skills required for Higher Biology with questions provided as examples. Try these questions then use the Problem Solving: Answers and Commentary chapter to mark your answers.

The **Writing Extended Responses** chapter offers guidance on this crucial area. This section should help you approach an area that is found difficult by many candidates.

The **Examination Tips** chapter has some helpful advice on how to tackle the paper in general terms and how to approach the different types of question. There is also guidance on timing.

The **Answers and Commentary** chapters have SQA standard answers to the Exam Questions and Problem Solving features with a useful commentary in *italic* type.

Top Tip

The top tip box is also used where we are offering a specific examination answering tip.

Brief Course Outline

The National Course *Higher Biology* is divided into **four** parts.

There are **three** National Units and **one** Course Examination.

A candidate needs to pass all **four** parts to gain a Course award.

The National Units are assessed at your school or college but the Course Examination is set, marked and graded by the SQA.

Each National Unit has a **Knowledge and Understanding** outcome **(O1)** and a set of **Problem Solving (O2)** and **Practical (O3)** outcomes to be mastered.

O1 The **Knowledge and Understanding** outcome is obviously different for each Unit. The Contents on page iii gives a breakdown of the Topic Areas into which each Unit is divided.

O2 & 3 The good news is – the **Problem Solving** and **Practical** outcomes are the **same** for each Unit! Our Problem Solving chapter on page 112 includes a list of the skills needed in these areas.

O3 Report – As well as passing the Unit tests (NABs) for each of the three Units, candidates submit a short report on an experiment they have carried out. This is usually done as part of the first Unit of study.

Examination Structure

The Higher Biology Examination is a single paper consisting of a booklet of questions in three sections.

Answers to most questions are written into the booklet and the exam is designed to take 2.5 hours.

Section A contains 30 multiple-choice questions: 20 questions test **Knowledge and Understanding** and 10 test **Problem Solving and Practical Abilities**.

Each question carries a 1 mark allocation and is answered on a special grid.

Section B contains structured questions with an allocation of 80 marks of which about 50 marks cover Knowledge and Understanding and 30 marks cover **Problem Solving and Practical Abilities**.

In section B there is always a question set in the context of **a practical situation** that is often unfamiliar and another question involving the **handling of related data**.

In our **Exam Questions**, we have given these questions the prefix SBQ (Section B Question).

Section C contains two extended response (essay type) questions testing the selection, organisation and presentation of knowledge.

The first question (1) is structured into parts, offers a choice between two titles and carries 10 marks.

The second question (2) is open-ended also offers a choice of two titles but carries 8 marks with an additional mark for coherence and an additional mark for relevance.

In our **Exam Questions**, we have given these questions the prefix SCQ (Section C Question).

A note on question levels

Questions can be graded as C-type, B-type or A-type questions depending on their level of complexity and demand.

In **C-type questions**, the material tested tends to be straightforward and usually involves basic recall of knowledge.

B-type questions are more difficult, often involving two answers for one mark.

A-type questions are more complex and often involve explanations or calculations with several steps. Two parts to a response are usually required for one mark. Candidates are often required to apply their knowledge to unfamiliar contexts and to combine knowledge from different Units.

In our **Exam Questions**, we have given a guide to the question levels involved.

Chapter 1

CELL BIOLOGY

Unit 1 is called 'Cell Biology'. It comprises the following topic areas:

1.1 Cell Structure in Relation to Function

1.2 Photosynthesis

1.3 Energy Release

1.4 Synthesis and Release of Protein

1.5 Cellular Response in Defence

1.1 Cell Structure in Relation to Function

Key Ideas

- ☐ 1 Cells in one type of tissue can vary in structure.
- ☐ 2 Cells in different types of tissue vary in structure.
- ☐ 3 Differences in cell structure depend on the functions the cells perform.
- ☐ 4 Cells can differ in their size, shape and the number and types of organelles they contain.
- ☐ 5 Some organisms have only one cell. These are called unicellular organisms and they contain all the organelles needed to carry out the processes required for survival.
- ☐ 6 The outer boundary of an animal cell is called the plasma membrane.
- ☐ 7 Plant cells have a cell wall outside their plasma membrane.
- ☐ 8 Substances must pass through the plasma membrane and cell wall to enter or leave a cell.
- ☐ 9 The movement of small molecules or ions of a substance in or out of a cell occurs by diffusion, osmosis or active transport.
- ☐ 10 Plant cell walls are fully permeable to all small molecules.
- ☐ 11 Plant cell walls are made of criss-cross layers of cellulose fibres.
- ☐ 12 Plant cell walls provide support and prevent cells from bursting when full of water.
- ☐ 13 Plasma membranes are selectively permeable. This means they let molecules of certain substances pass through.
- ☐ 14 Plasma membranes are made of protein and phospholipid arranged as a fluid mosaic.
- ☐ 15 The phospholipid molecules of the plasma membrane are in a constantly moving double layer allowing the membrane the flexibility to change shape. This is why the membrane is described as fluid.
- ☐ 16 The proteins are scattered through the phospholipid double layer giving a mosaic pattern.

***Key Ideas** continued* ➢

Key Ideas continued

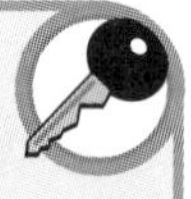

- ☐ 17 Some proteins are on the surfaces on the phospholipid layer, others extend across the phospholipid layer and some form pores passing across the membrane.
- ☐ 18 Pores in the membrane allow only small soluble molecules to pass freely across.
- ☐ 19 Damage to the membrane can cause loss of selective permeability. High temperatures can denature protein and treatment with alcohol can dissolve the phospholipid layer.
- ☐ 20 Diffusion is the movement of molecules from an area of high concentration to an area of lower concentration that results in the even spread of molecules.
- ☐ 21 Diffusion relies on the molecules' own movement energy and does not require further added energy.
- ☐ 22 Dissolved foods, oxygen and carbon dioxide are examples of substances that move into and out of cells by diffusion.
- ☐ 23 Osmosis is a special type of diffusion.
- ☐ 24 Osmosis is the diffusion of water molecules from a region of high water concentration to a region of lower water concentration through a selectively permeable membrane.
- ☐ 25 Hypotonic solutions are higher in water concentration compared to others.
- ☐ 26 Hypertonic solutions are lower in water concentration compared to others.
- ☐ 27 Isotonic solutions are equal in water concentration.
- ☐ 28 Animal cells will take up water, swell and burst if placed in a hypotonic solution.
- ☐ 29 Animal cells will lose water and tend to shrink if placed in a hypertonic solution.
- ☐ 30 Plant cells will take up water, their vacuoles will swell and the cells become turgid in hypotonic solutions.
- ☐ 31 In hypertonic solutions, plant cells will lose water; the vacuoles will shrink and pull the plasma membrane and cytoplasm from the cell walls. The cells become flaccid and plasmolysed.
- ☐ 32 In isotonic solutions, neither animal nor plant cells experience net movement of water.
- ☐ 33 Active transport is the movement of substances from low to high concentration against the concentration gradient.
- ☐ 34 Active transport requires additional energy.
- ☐ 35 ATP produced in respiration supplies the additional energy required for active transport.
- ☐ 36 Specialised proteins in the plasma membrane act as carriers in active transport. They can recognise specific molecules and transport them across the plasma membrane.
- ☐ 37 Active transport allows for selective ion uptake that results in the accumulation of specific ions in certain cells.

Topic Notes

Cell variety

Cells can vary in their structure within the same tissue. In blood, for example, there are several types of cell that have different structures depending on their functions.

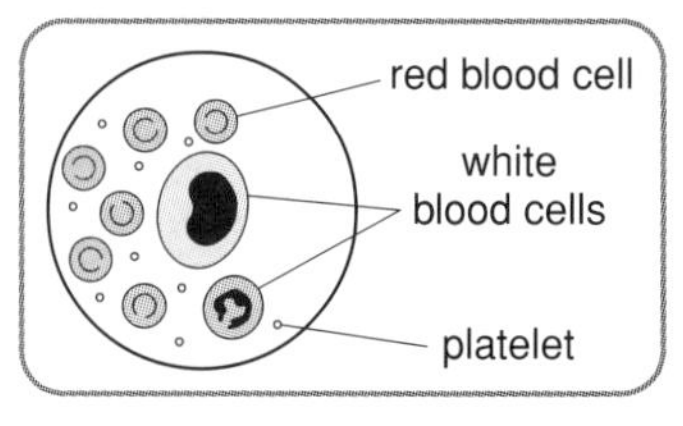

Figure 1.1 Cells in blood tissue

Cells also vary from one tissue to another. For example, the cells from palisade mesophyll differ from the cells in spongy mesophyll.

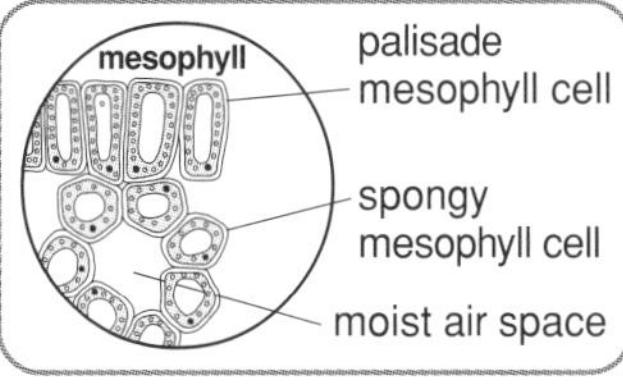

Figure 1.2 Cells in palisade and spongy mesophyll of a leaf

Top Tip

The next point is really needed for full understanding of the idea of cell structure related to function although the content is drawn from other sections of this book.

When cells are examined closely using powerful microscopes, structures called organelles become visible. The organelles are the sites of specialised functions. For example, chloroplasts carry out photosynthesis. Cells vary in the types and numbers of organelles that they contain. The table gives information about various organelles and the functions they perform.

Organelle	Functions
Nucleus	Contains genetic information to control all cell activities
Chloroplast	Site of photosynthesis
Mitochondrion	Site of aerobic respiration
Ribosome	Site of protein synthesis
Rough endoplasmic reticulum	Transport of protein
Golgi apparatus	Processes and packages protein for secretion
Lysosome	Destroys unwanted material using digestive enzymes
Centriole	Produces the spindle fibres in mitosis and meiosis

The differences in cell structure both within and between tissues are related to the functions of the cells involved. The table gives information about the structure and function of cells you studied at Standard Grade or in Intermediate 2 Biology.

Type of cell	Structure	Structural Features	Function
Root hair See 2.3		Hair shape provides an increased surface area	Absorption of water and minerals

Type of cell	Structure	Structural Features	Function
Phloem See 3.1	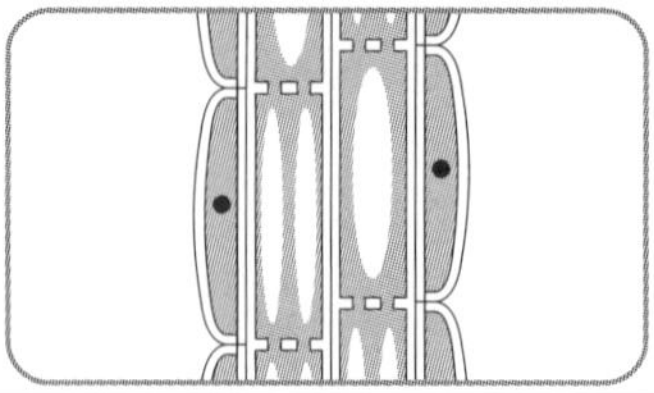	Cells have end wall pores allowing cytoplasm to connect	Transport of products of photosynthesis
Xylem See 3.1	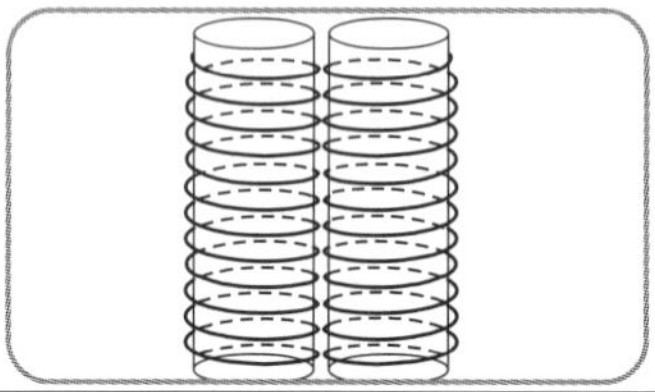	Cells are dead, thickened with lignin and with no end walls or contents	Transport of water and minerals Support of plant
Palisade mesophyll See 1.2	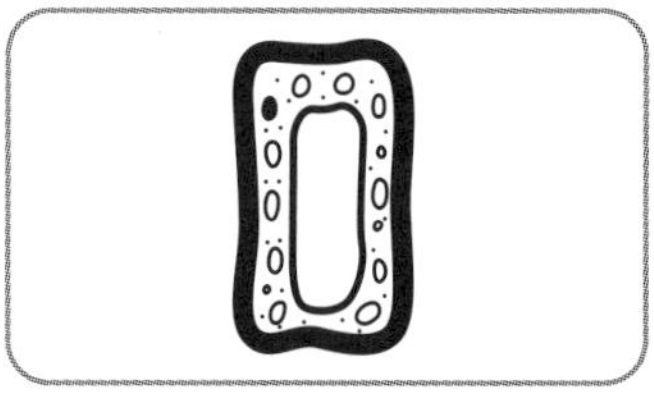	Contain many chloroplasts	Photosynthesis
Skeletal muscle See 1.3	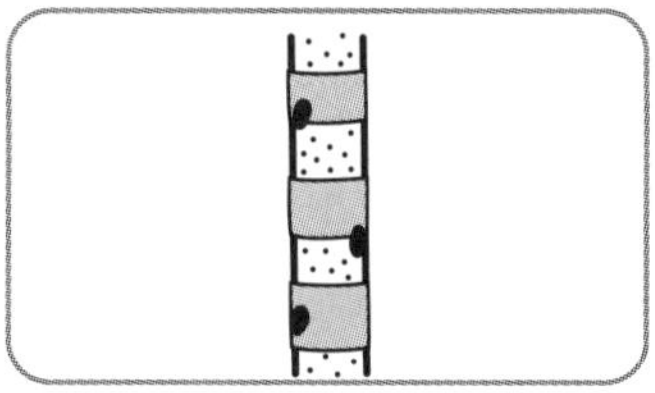	Contain many mitochondria to provide ATP for muscle contraction	Movement
Goblet cell in trachea See 1.4	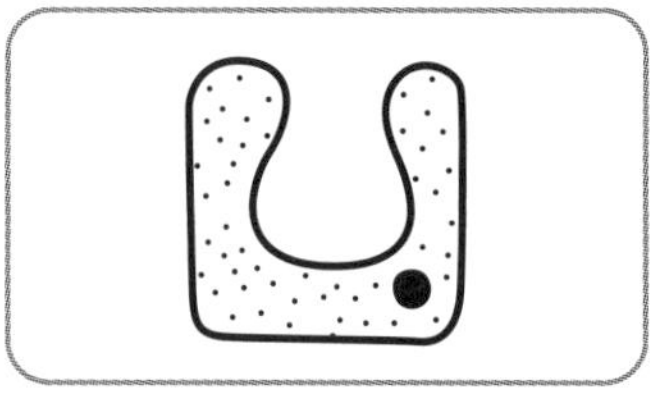	Contain many Golgi	Secrete mucus to trap bacteria and other breathed-in particles
Red blood cell	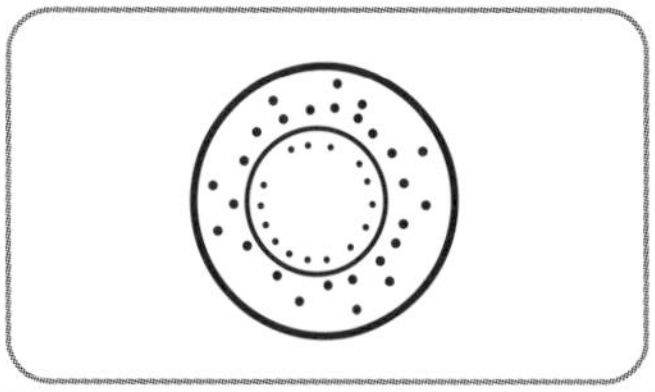	Disc-shaped with dimples in each side to provide increased surface area	Absorb oxygen from breathed-in air and release it to cells
Phagocyte See 1.5	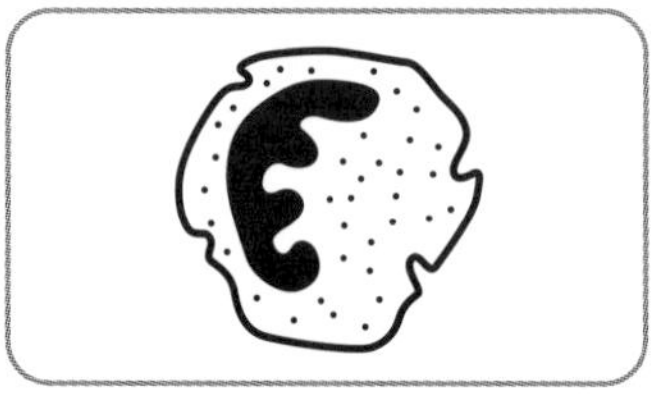	Have many lysosomes containing digestive enzymes	Engulf and digest invading bacteria

Unicellular organisms have only one cell and contain the organelles they need to survive. *Euglena* is an example of a unicellular organism and the diagram describes how its structure is related to survival.

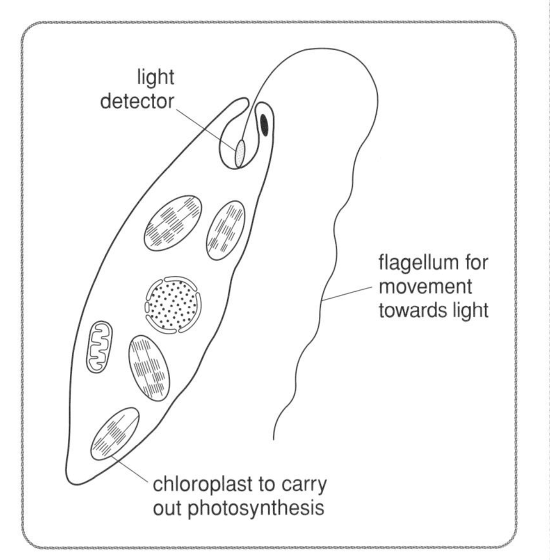

Figure 1.3 *Euglena*

Absorption and secretion of materials

Animal cells are surrounded by a plasma membrane and plant cells have a plasma membrane that is surrounded by a cell wall. The diagram shows the position of these structures.

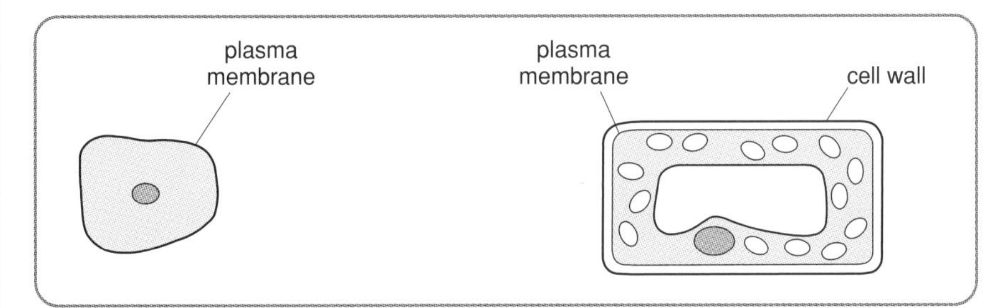

Figure 1.4 Animal cell (left) and plant cell (right)

The molecules of substances entering or leaving cells must pass through these boundaries. The cell wall is fully permeable and does not provide a barrier to the passage of small dissolved molecules. It is made up of cellulose fibres in a criss-cross arrangement with large gaps. The cell wall provides support for plant cells and helps to prevent it busting when full of water.

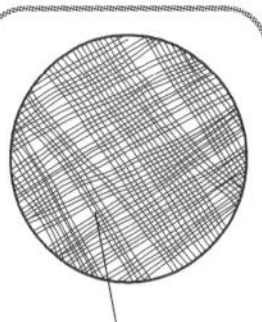

Figure 1.5 Arrangement of fibres in a cell wall

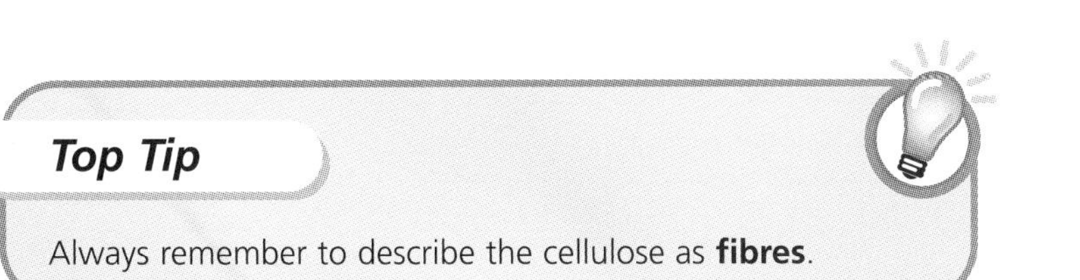

Top Tip

Always remember to describe the cellulose as **fibres**.

The plasma membrane is composed of protein and phospholipid in an arrangement called a fluid mosaic. The phospholipid molecules form a double layer and are in constant motion giving the fluid nature to the membrane and allowing it great flexibility.

Figure 1.6 shows the main features of the fluid mosaic model of the plasma membrane.

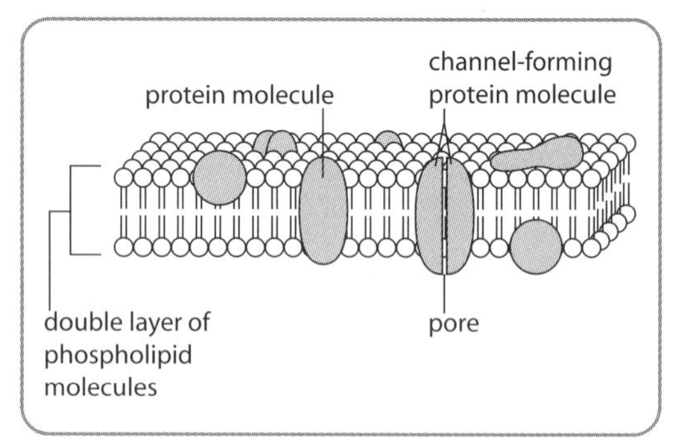

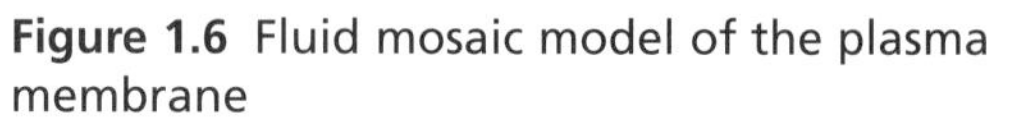

Figure 1.6 Fluid mosaic model of the plasma membrane

Top Tip

The 3Fs! – Phospholipid – **F**luid – **F**lexible

In your exam, you must use the term phospholipid – **not** lipid or fat!

The proteins are scattered within the phospholipid layer in a patchy mosaic pattern. Some proteins are attached to the surfaces of the phospholipid layer and some are embedded within it. Others contain pores that penetrate across the membrane.

Top Tip

The 3Ps – **P**rotein – **P**atchy – **P**orous

The pores allow the free passage of small soluble molecules through the membrane and give the membrane its selective permeability.

Treatment with heat can denature the protein causing it to lose its selective permeability and become freely permeable. Substances such as alcohol can dissolve the phospholipid layer with the same result.

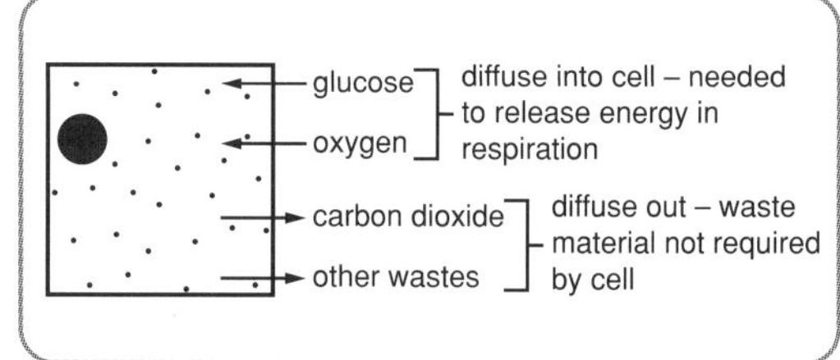

Figure 1.7 Examples of substances entering and leaving a cell by diffusion

Diffusion is the movement of small soluble molecules from high concentration to lower concentration leading to an even spread. The energy required for this movement comes from the movement energy of the molecules diffusing. No additional energy is required. Dissolved foods such as glucose and amino acids, and the gases oxygen and carbon dioxide are examples of substances that can diffuse into and out of cells.

Top Tips

The **2Ds** substances must be **d**issolved before they can **d**iffuse.

If you are asked to **explain** the importance of diffusion to living cells, you should state that the cells obtain glucose and oxygen by diffusion for aerobic respiration to provide ATP.

Osmosis is a special type of diffusion. It describes the movement of water molecules from a region of high water concentration to a region of lower water concentration through a selectively permeable membrane.

Concentrations of solutions are usually described in molarity (M) or sometimes as percentages (%).

Hypotonic solutions contain higher water concentrations than others. Fresh water is hypotonic to the cytoplasm of living cells. The most hypotonic solutions have the lowest molarities or the lowest % of dissolved solutes. Cells take in water from hypotonic solutions.

Animal cells gain water by osmosis from hypotonic solutions, swell and burst. Plant cells also gain water from hypotonic solutions and although they swell a little, their walls prevent bursting. Plant cells swollen with water are said to be turgid.

Hypertonic solutions contain lower water concentrations than others. Seawater is hypertonic to the cytoplasm of living cells. The most hypertonic solutions have the highest molarities or highest percentage of dissolved solutes. Cells lose water to hypertonic solutions.

Animal cells lose water by osmosis to hypertonic solutions and shrink. Plant cells also lose water to hypertonic solutions causing their cell contents to shrink, leaving the cell plasmolysed and flaccid.

Solutions that are isotonic have the same water concentration as each other. The plasma of blood is usually isotonic to the cytoplasm of the blood cells within it. There is no net gain or loss of water to cells in isotonic solutions.

Active transport involves the movement of molecules from low to high concentration. This type of movement needs additional energy that is available from ATP. Cells must be respiring aerobically to produce enough ATP for active transport. It involves specialised proteins in the plasma membrane acting as carriers to pass the molecules across.

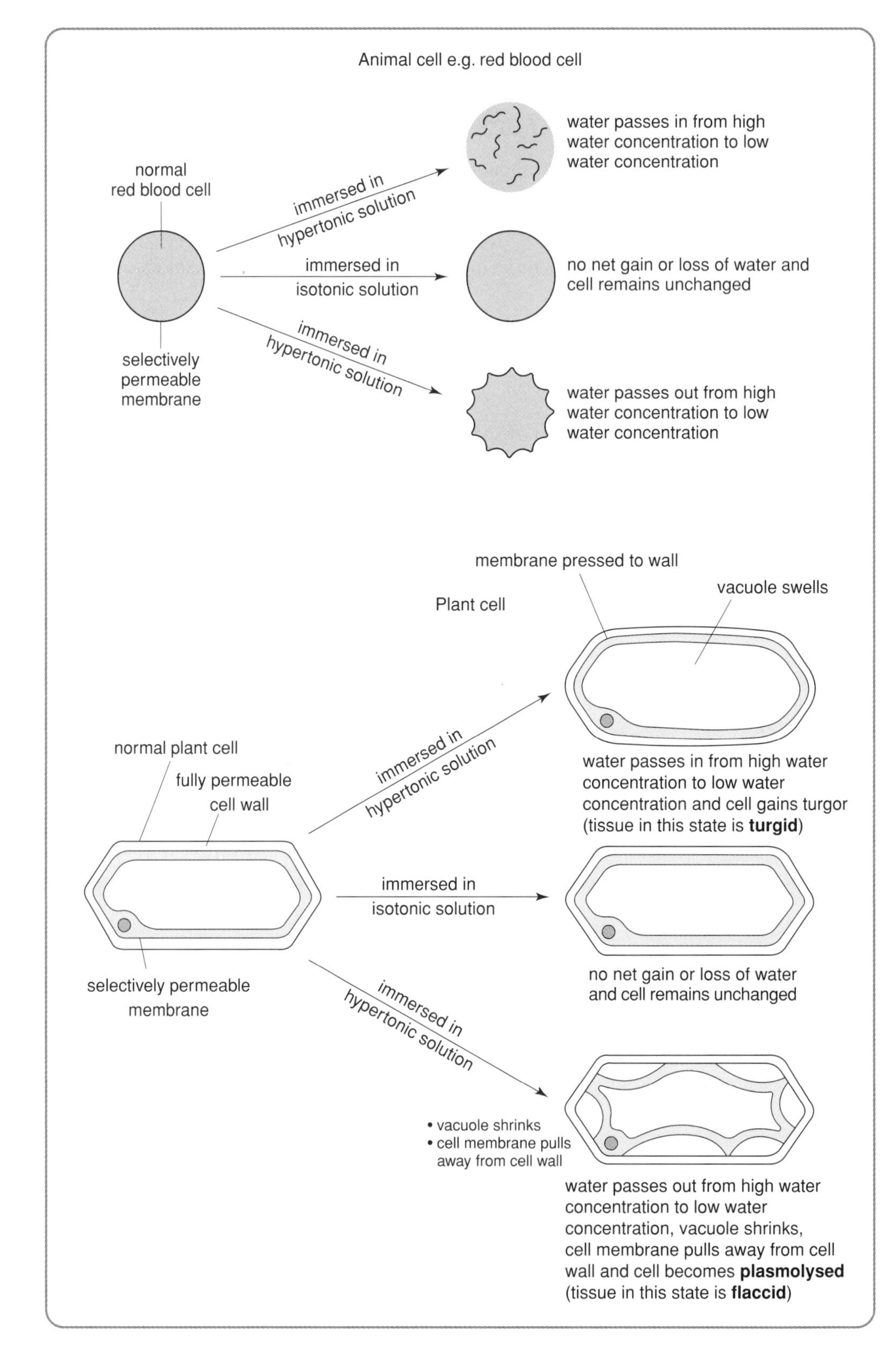

Figure 1.8 The effects of osmosis on animal and plant cells

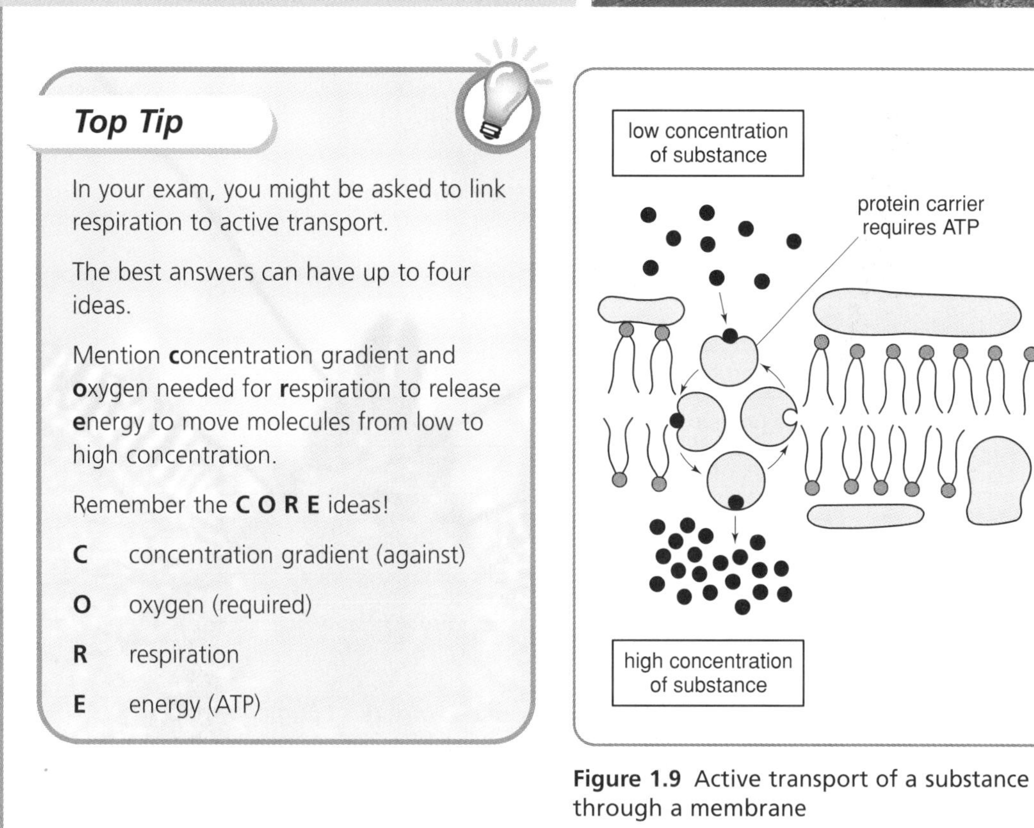

Top Tip

In your exam, you might be asked to link respiration to active transport.

The best answers can have up to four ideas.

Mention **c**oncentration gradient and **o**xygen needed for **r**espiration to release **e**nergy to move molecules from low to high concentration.

Remember the **C O R E** ideas!

C concentration gradient (against)

O oxygen (required)

R respiration

E energy (ATP)

Figure 1.9 Active transport of a substance through a membrane

Exam Questions

Marks

SBQ 1 The diagrams below show some of the structures present in animal and plant cells.

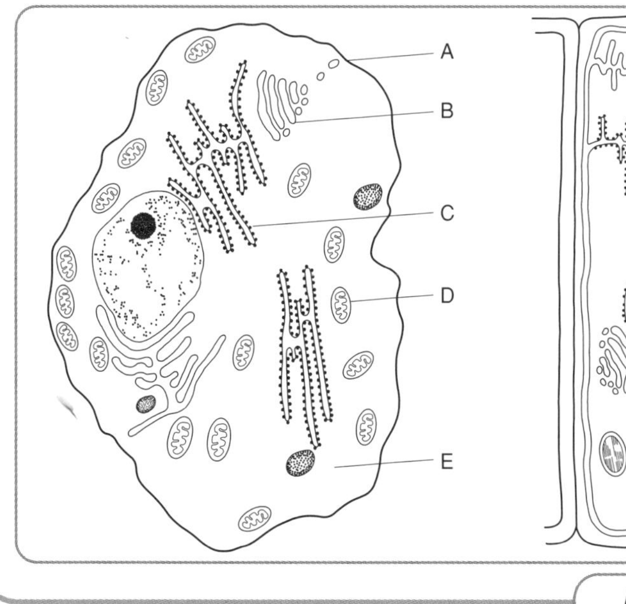

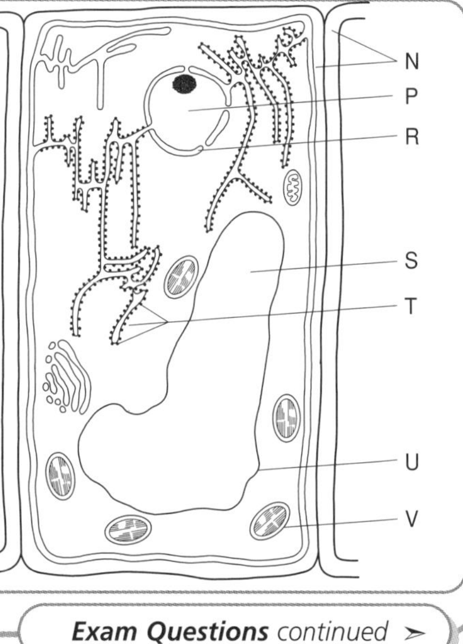

Exam Questions *continued* ➢

Exam Questions *continued*

Complete the table by entering letters from the diagrams that match each function.

Function	Letter
Site of photosynthesis	
Site of aerobic respiration	
Site of protein synthesis	
Transport of proteins	
Controls the movements of materials into and out of the cell	
Processing and packaging of secretions	

3C

SBQ 2 The diagram below shows the unicellular organism *Euglena* that is found in ponds.

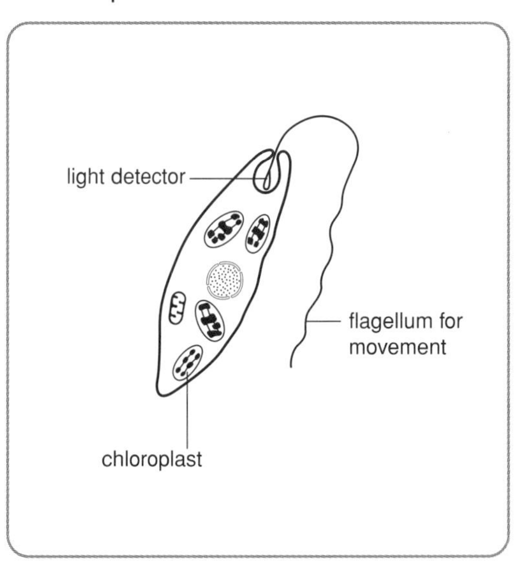

Use information from the diagram to explain how the structure of *Euglena* is related to the efficiency of photosynthesis. 2C/B

SBQ 3 The table below contains descriptions that refer to either the cell wall or the plasma membrane.

Exam Questions *continued* ➢

Exam Questions continued

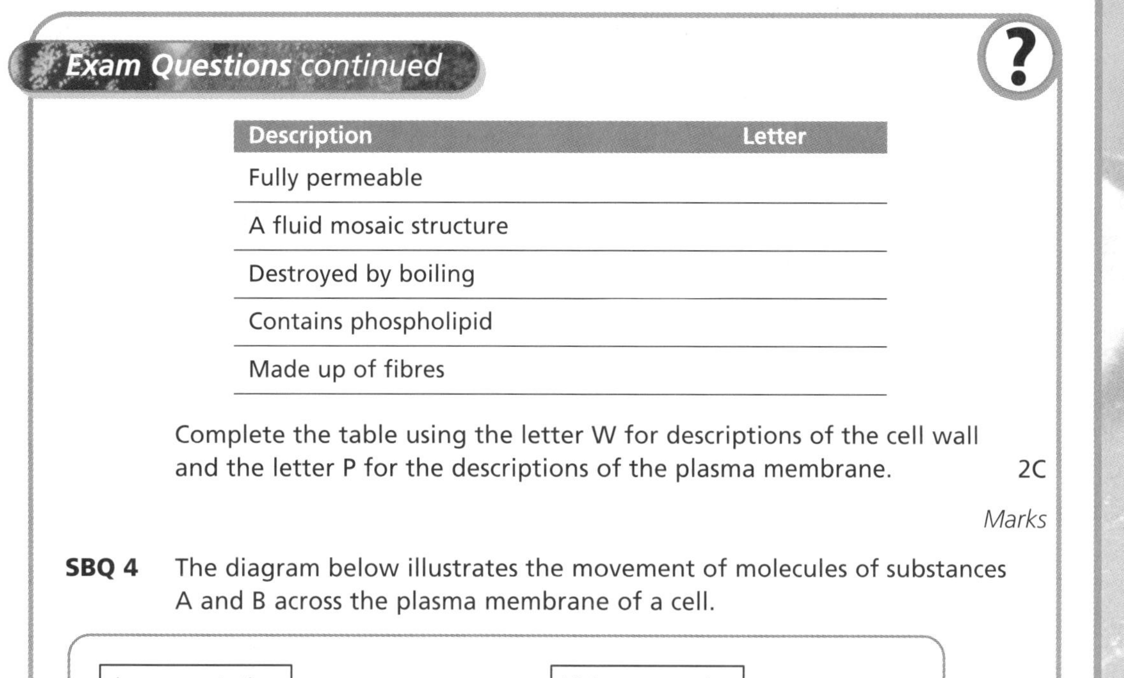

Description	Letter
Fully permeable	
A fluid mosaic structure	
Destroyed by boiling	
Contains phospholipid	
Made up of fibres	

Complete the table using the letter W for descriptions of the cell wall and the letter P for the descriptions of the plasma membrane. 2C

Marks

SBQ 4 The diagram below illustrates the movement of molecules of substances A and B across the plasma membrane of a cell.

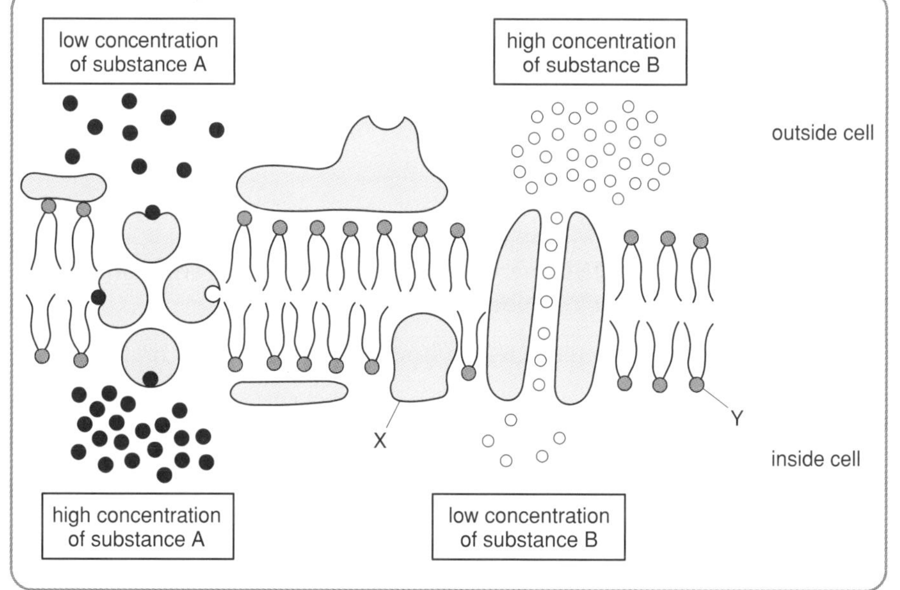

(a) Name parts X and Y. 1C

(b) Identify the methods by which substances A and B enter the cell. 1B

SBQ 5 The graph below compares the uptake of nitrate ions by barley roots in the presence and absence of oxygen.

Exam Questions *continued* ➢

Exam Questions *continued*

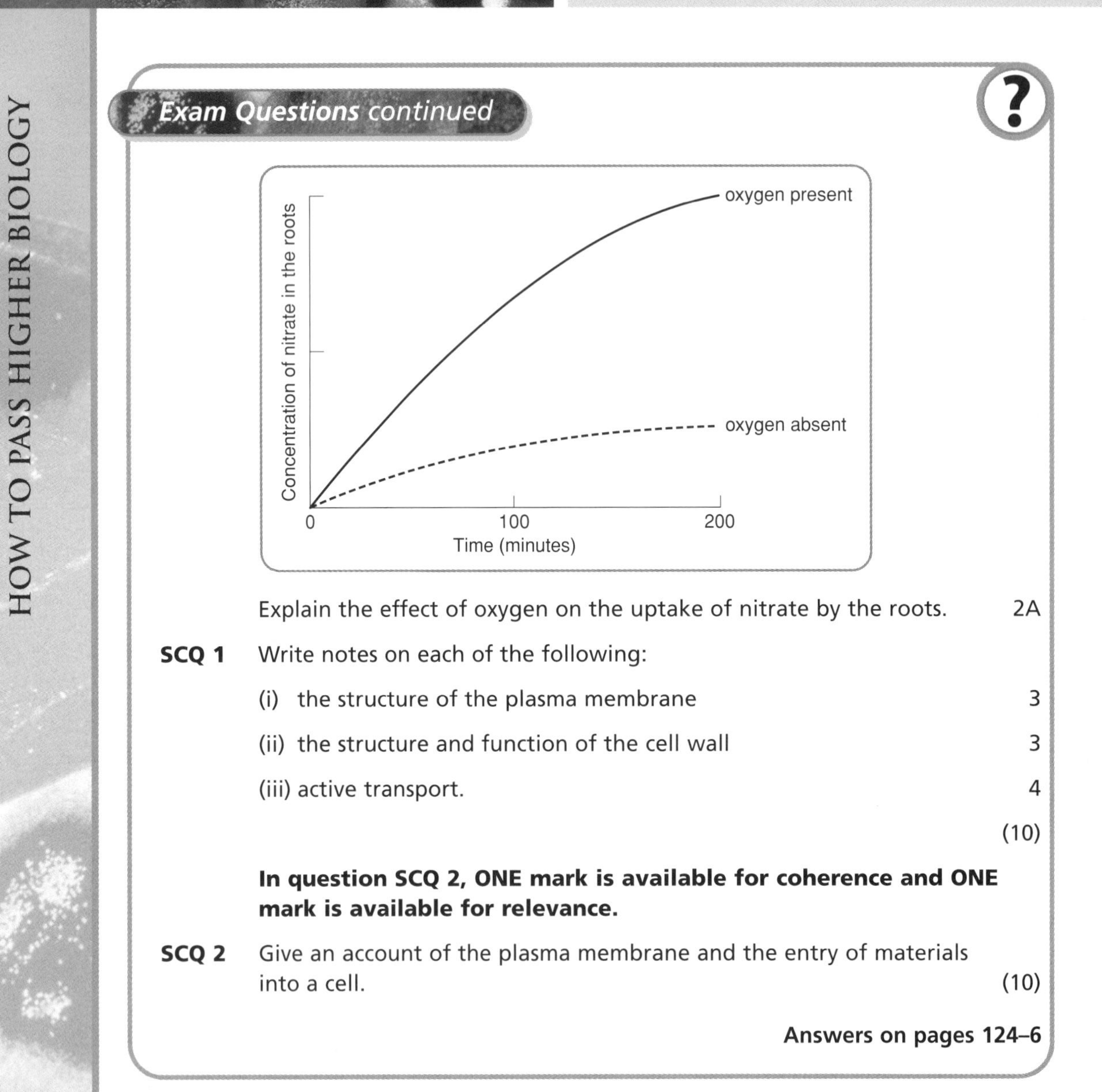

Explain the effect of oxygen on the uptake of nitrate by the roots. 2A

SCQ 1 Write notes on each of the following:

(i) the structure of the plasma membrane 3

(ii) the structure and function of the cell wall 3

(iii) active transport. 4

(10)

In question SCQ 2, ONE mark is available for coherence and ONE mark is available for relevance.

SCQ 2 Give an account of the plasma membrane and the entry of materials into a cell. (10)

Answers on pages 124–6

1.2 Photosynthesis

Key Ideas

- ☐ 1 Rays of visible light are made up of a range of wavelengths. The different wavelengths have different colours – red, orange, yellow, green, blue, indigo and violet – which make up the spectrum of visible light.
- ☐ 2 The light that strikes the surface of a leaf can be absorbed, reflected or transmitted.
- ☐ 3 Photosynthetic pigments in the leaves absorb light energy and change it into chemical energy.
- ☐ 4 The main photosynthetic pigment is chlorophyll.

Key Ideas *continued* ➢

Key Ideas continued

- ☐ 5 The wavelengths of light that are absorbed by a pigment are called its absorption spectrum.
- ☐ 6 Absorption of light by chlorophyll occurs mainly in the blue and red regions of the spectrum.
- ☐ 7 Accessory pigments absorb some light from other regions of the spectrum and pass the energy on to chlorophyll.
- ☐ 8 The accessory pigments include carotene and xanthophyll.
- ☐ 9 The wavelengths of light actually used by a pigment in photosynthesis are called its action spectrum.
- ☐ 10 Some of the light energy is used in the regeneration of ATP and also in the splitting of water.
- ☐ 11 The photosynthetic pigments of a leaf can be extracted from the leaf and separated by means of chromatography.
- ☐ 12 Photosynthesis occurs in the chloroplasts of plant cells.
- ☐ 13 Chloroplasts are bounded by a double membrane. They contain stacks of membranes called grana surrounded by a fluid region called the stroma.
- ☐ 14 Photosynthesis occurs in two stages: the light dependent stage and the carbon fixation stage.
- ☐ 15 The light dependent stage occurs in the grana that contain the photosynthetic pigments.
- ☐ 16 Some light energy absorbed by pigments in the grana is used to produce ATP for use in the carbon fixation stage.
- ☐ 17 Some light energy is used in photolysis. This is the splitting of water releasing oxygen as a by-product and hydrogen.
- ☐ 18 The hydrogen is transferred to the carbon fixation stage by the hydrogen acceptor NADP that becomes reduced to form NADPH.
- ☐ 19 The ATP and the NADPH from the light dependent stage are transferred to the carbon fixation stage.
- ☐ 20 The carbon fixation stage occurs in the stroma of chloroplasts.
- ☐ 21 The carbon fixation stage involves the production of glucose, a carbohydrate, by means of an enzyme-controlled sequence of reactions, also called the Calvin cycle.
- ☐ 22 The Calvin cycle requires energy provided by ATP from the light dependent stage.
- ☐ 23 In the Calvin cycle, carbon dioxide is reduced to carbohydrate in the form of glucose using hydrogen provided by the light dependent stage.
- ☐ 24 The carbon dioxide enters the Calvin cycle where it is accepted by the 5-carbon compound ribulose 1,5 biphosphate (RuBP).
- ☐ 25 Having accepted the carbon dioxide, the RuBP forms a 6-carbon compound, which immediately splits into two molecules of the 3-carbon compound glycerate 3-phosphate (GP).
- ☐ 26 GP accepts the hydrogen from the light dependent stage and is reduced to carbohydrate in the form of glucose. This reaction also requires energy from ATP.
- ☐ 27 Some of the GP is converted back to RuBP and so continues the cycle.
- ☐ 28 Major biological molecules in plants such as proteins, fats, carbohydrates and nucleic acids are derived from the photosynthetic process.

Topic Notes

The role of light and photosynthetic pigments

Photosynthesis is simply the way green plants make their own foods such as glucose. They use light energy absorbed by coloured pigments, such as chlorophyll, and the raw materials water and carbon dioxide from the environment.

Top Tip

Learn this simple equation for photosynthesis!

$$\text{water} + \text{carbon dioxide} \xrightarrow[\text{coloured pigment}]{\text{light energy}} \text{glucose} + \text{oxygen}$$

It is worth noting that the photosynthesis equation can be balanced. This helps in learning the carbon numbers involved since these need to be known for your examination.

The balanced equation is given below.

$6CO_2$ (carbon dioxide) + $6H_2O$ (water) $\longrightarrow$ $C_6H_{12}0_6$ (glucose) + 60_2 (oxygen)

This simply means that six molecules of carbon dioxide and six molecules of water are needed to make one glucose molecule and there are six molecules of oxygen produced as a by-product.

Absorption, transmission and reflection of light by a leaf

Most of the light that strikes a leaf is absorbed but very little of the energy is actually converted into useful chemical energy. Much of the absorbed light is lost as heat. The rest of the light striking the leaf is either reflected off the leaf surface or is transmitted through the leaf.

Top Tip

Remember **ART**! These processes are sometimes called the **fates** of light striking a leaf.

Absorption	Light is taken into the leaf cells.
Reflection	Light is bounced back from the leaf surface.
Transmission	Light passes right through the leaf.

Role of chlorophyll and other photosynthetic pigments

The photosynthetic pigments are found in the grana of the chloroplasts. They absorb light energy and convert it into chemical energy. Chlorophyll a is the main photosynthetic pigment. It absorbs mainly blue and red light wavelengths. Chlorophyll b, carotene and xanthophyll are accessory pigments that absorb other wavelengths of light and pass the energy on to chlorophyll a. The accessory pigments broaden the absorption spectrum so that more energy is available for photosynthesis.

Top Tip

The spectrum of light can be seen if a beam is shone through a glass prism onto a screen. The spectrum is a rainbow of colours of different wavelengths.

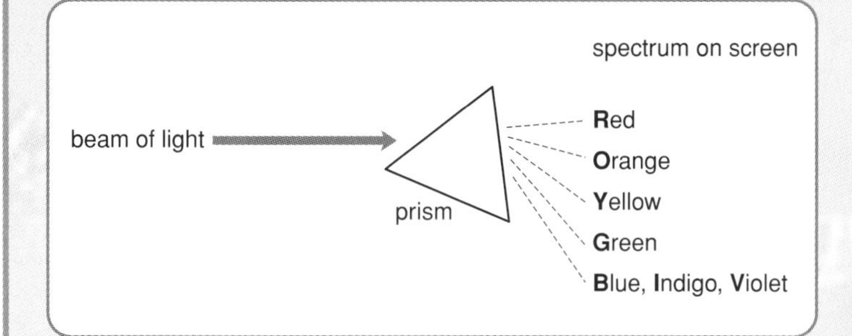

Figure 1.10 Light split into its spectrum by a prism

Richard **O**f **Y**ork **G**ave **B**attle **I**n **V**ain – you can use this memory aid to help you remember the order of the colours.

An absorption spectrum shows the extent to which different colours of light are absorbed by a pigment. This can be shown as a graph.

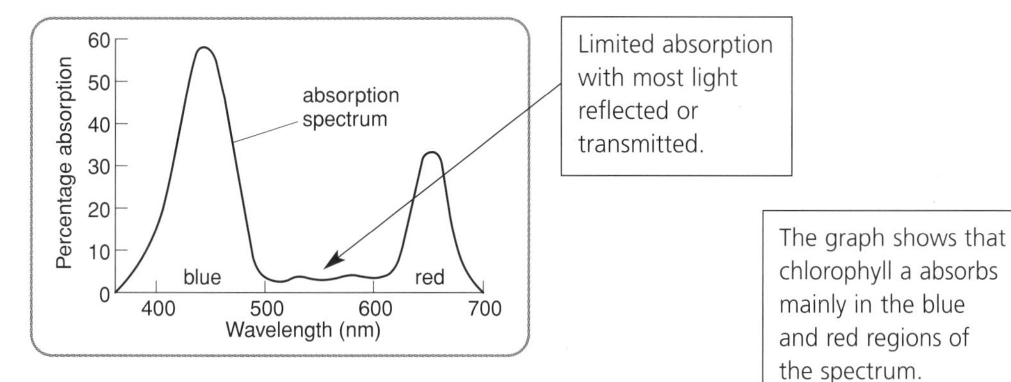

Figure 1.11 The absorption spectrum for chlorophyll a

An action spectrum shows the rate of photosynthesis carried out in different colours of light. Again, this can be shown as a graph.

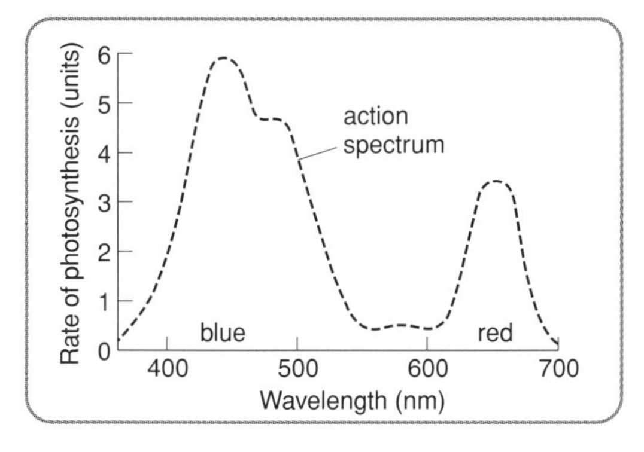

The graph shows that the rate of photosynthesis is highest in the blue and red regions of the spectrum.

Figure 1.12 The action spectrum for a mixture of pigments from a green plant leaf

Top Tip

If the action spectrum is placed on top of the absorption spectrum for chlorophyll a, you can see that the two do not **exactly** match.

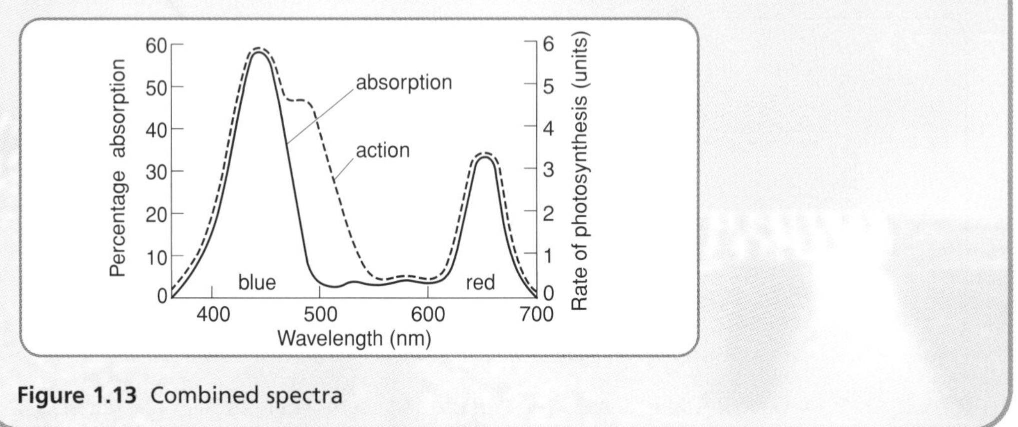

Figure 1.13 Combined spectra

Photosynthesis is still effective in colours of light **not** well absorbed by chlorophyll a. This provides evidence that chlorophyll a is not the **only** pigment involved, and suggests that the accessory pigments must be involved in absorbing other wavelengths of light being used in photosynthesis.

Separation of photosynthetic pigments by means of chromatography

For your examination, you are expected to know how to extract the pigments from a leaf and separate pigments by chromatography.

Paper chromatography involves the following steps:

- Grind leaves with acetone to extract the pigments
- Filter to remove cell debris from the extract
- Repeatedly apply the extract to form a concentrated spot on a strip of chromatography paper.
- Place the paper into a tube of solvent with the spot above the solvent surface.
- Allow time for the solvent to run and separate the pigments.
- Pigments with higher solubility will run further up the paper.

The order of separation with the most soluble first is shown in Figure 1.14.

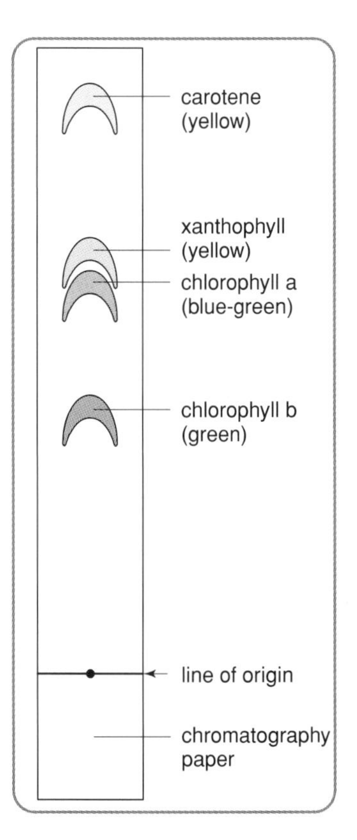

Figure 1.14 Separation of pigments in a leaf, using chromatography

Top Tip

Remember the order of pigments from the bottom as **bax c.**

The stages of photosynthesis occur inside chloroplasts.

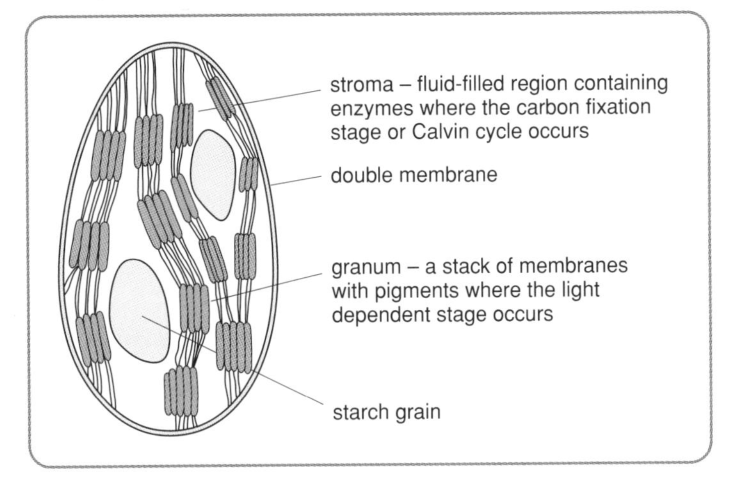

Figure 1.15 Structure of a chloroplast

Top Tip

A simple, but well-labelled, diagram like this can earn you useful marks in extended response questions.

The light dependent and carbon fixation stages

Learn photosynthesis in two stages!

Stage 1 – the light dependent stage

These are the main features of the light dependent stage you need to remember for the exam:

- Happens in the grana of chloroplasts.
- Light energy is absorbed by pigments.
- Some of the light energy is used to split water to obtain hydrogen. This is called photolysis.
- The hydrogen combines with the hydrogen acceptor NADP to form NADPH.
- ATP is produced using some of the light energy that has been absorbed.
- Photolysis also releases the by-product oxygen.

Top Tip

NAD**P** is the hydrogen acceptor in **P**hotosynthesis – don't get it confused with NAD, a carrier in respiration!

Figure 1.16 shows the light dependent stage of photosynthesis.

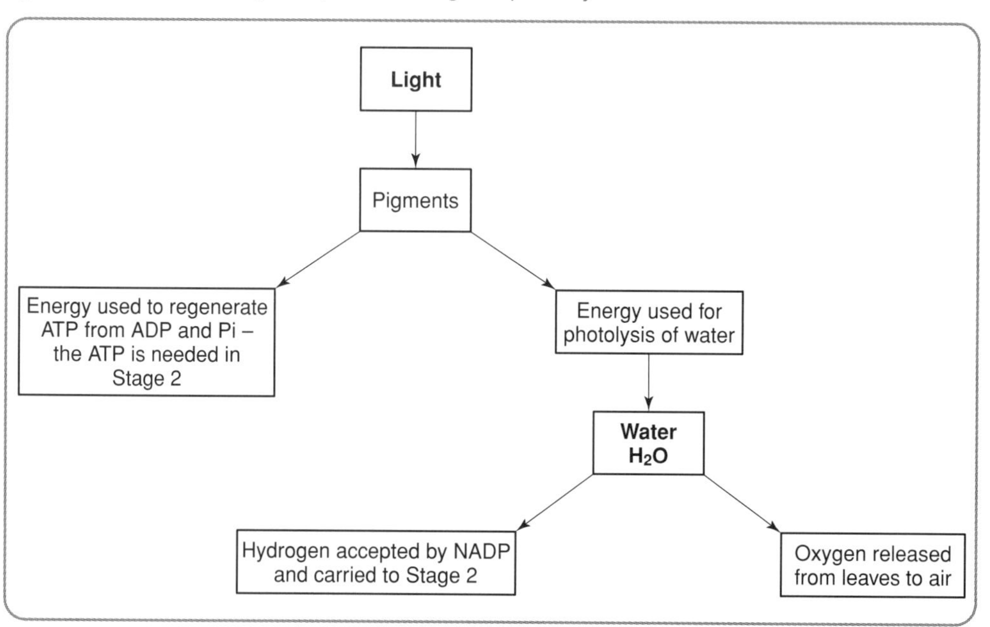

Figure 1.16 Light dependent stage of photosynthesis

The ATP and NADPH produced in the light dependent stage are then transferred from the grana to the stroma for use in the carbon fixation stage.

Stage 2 – the carbon fixation stage

These are the main features of the carbon fixation stage (Calvin cycle) you need to remember for your exam:

- Happens in the stroma of the chloroplasts.
- Involves an enzyme-controlled sequence of reactions.
- Requires ATP produced in the light dependent stage.
- Requires hydrogen from the NADPH produced in the light dependent stage.
- Carbon dioxide is accepted by 5-carbon RuBP.
- Two molecules of 3-carbon GP are produced.
- The ATP and hydrogen are used to reduce the GP to produce carbohydrate in the form of glucose.
- Some GP is used to regenerate RuBP.

Figure 1.17 shows the carbon fixation stage (Calvin cycle).

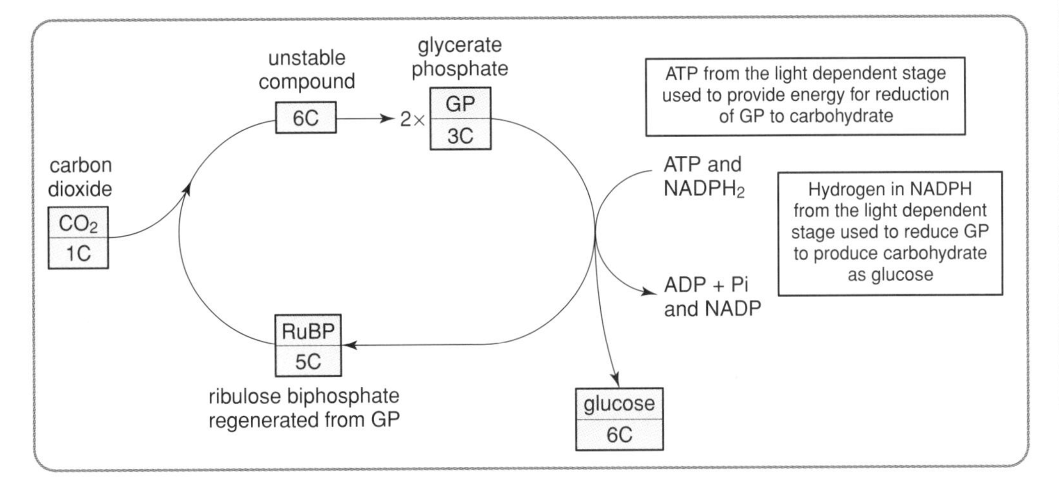

Figure 1.17 Carbon fixation stage (Calvin cycle)

Top Tip

The carbon fixation stage essentially involves using ATP and hydrogen from the light dependent stage to reduce carbon dioxide to carbohydrate.

The equation in the diagram summarises the stages of photosynthesis.

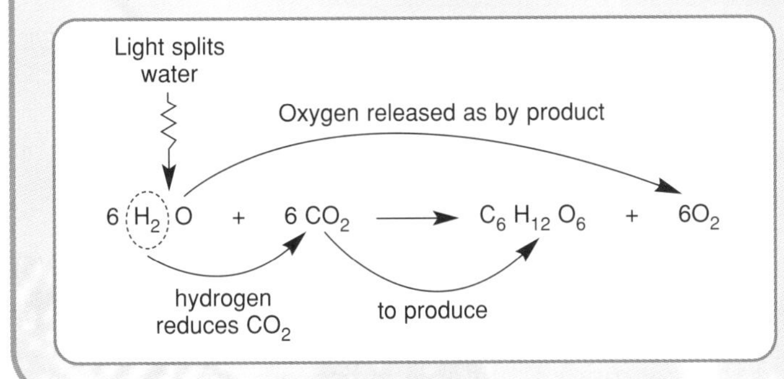

Top Tip

You must know the numbers of carbon atoms per molecule of the substances in the Calvin cycle. Try memorising these as a table.

Substance	Carbon atoms per molecule
Carbon dioxide	1
GP	3
RuBP	5
Glucose	6

The major biological molecules in plants, proteins, fats, carbohydrates and nucleic acids are **all** derived from the photosynthetic process.

Exam Questions

Marks

SBQ 1 (a) Give two possible fates of light that shines on the leaves of plants. 2C

(b) State the region of a chloroplast in which light is absorbed. 1C

(c) Name the two products of the light dependent stage which are required for the conversion of GP into glucose. 1C

(d) Within which region of the chloroplast does the conversion of GP to RuBP occur? 1C

(e) Complete the table by inserting the number of carbon atoms found in each molecule of the substances.

Substance	Number of carbon atoms in each molecule
Carbon dioxide	
GP	
RuBP	
Glucose	

2C/B

SCQ 1A Give an account of photosynthesis under the following headings:

(i) structure of a chloroplast 4

(ii) light dependent stage of photosynthesis. 6

(10)

SCQ 1B Write notes on photosynthesis under the following headings:

(i) the extraction and separation of photosynthetic pigments by chromatography 4

(ii) the carbon fixation stage (Calvin cycle). 6

(10)

In SCQ 2, ONE mark is available for coherence and ONE mark is available for relevance.

SCQ 2 Give an account of chloroplast structure in relation to the location of the stages of photosynthesis and describe the extraction and separation of photosynthetic pigments by chromatography.

Answers on page 126–8

1.3 Energy Release

Key Ideas

- ☐ 1 ATP is a substance that can transfer chemical energy from a respiratory substrate, such as glucose, to energy-requiring processes in cells.
- ☐ 2 ATP provides the energy for processes such as muscle contraction, active transport, DNA replication and protein synthesis.
- ☐ 3 When ATP releases its energy, it breaks down into adenosine diphosphate (ADP) and inorganic phosphate (Pi).
- ☐ 4 ATP is regenerated as adenosine diphosphate (ADP) and inorganic phosphate (Pi) combine using energy released from a respiratory substrate.
- ☐ 5 Respiration is a series of enzyme-controlled reactions in which a respiratory substrate is oxidised to form carbon dioxide.
- ☐ 6 Oxidation involves the removal of hydrogen from a respiratory substrate.
- ☐ 7 The respiratory substrate is usually glucose, a sugar with 6 carbon atoms per molecule.
- ☐ 8 Proteins and fats are alternative respiratory substrates.
- ☐ 9 Respiration is accompanied by the synthesis of ATP from ADP and Pi.
- ☐ 10 The first stage in the respiration of glucose is called glycolysis.
- ☐ 11 Glycolysis occurs in the cytoplasm of cells.
- ☐ 12 Glycolysis does not require oxygen and involves the step-by-step breakdown of 6-carbon glucose molecules to form two 3-carbon pyruvic acid molecules.
- ☐ 13 Glycolysis is accompanied by a net gain of 2 ATP molecules. It is a **net** gain as some ATP is used in starting glycolysis but more is produced.
- ☐ 14 Glycolysis also releases hydrogen which is carried away by NAD in the form NADH.
- ☐ 15 The second stage in respiration is Krebs cycle that takes place within the central matrix of mitochondria.
- ☐ 16 Krebs cycle starts the aerobic phase of respiration and results in the production of carbon dioxide and hydrogen.
- ☐ 17 In Krebs cycle, each pyruvic acid molecule from glycolysis is converted to a 2-carbon acetyl group.
- ☐ 18 Each acetyl group joins with a co-enzyme A molecule (CoA) to form acetyl-CoA.
- ☐ 19 Acetyl-CoA reacts with a 4-carbon compound to form a 6-carbon compound called citric acid.
- ☐ 20 Citric acid is gradually converted back to the 4-carbon compound in a cyclical series of reactions – carbons being lost as carbon dioxide.
- ☐ 21 Krebs cycle produces hydrogen that is transferred to the last stage of respiration by NAD in the form of NADH.
- ☐ 22 The last stage of respiration is the cytochrome system that is located on the cristae of the mitochondria and consists of a series of carriers.
- ☐ 23 The cytochrome system is part of the aerobic phase of respiration. The hydrogen produced by oxidation reactions is finally received by oxygen to form water.
- ☐ 24 The hydrogen is passed through a series of carriers before being passed to oxygen as the final acceptor.
- ☐ 25 If oxygen is not present, hydrogen cannot pass through the system and complete oxidation cannot occur.

Key Ideas *continued* ➢

***Key Ideas** continued*

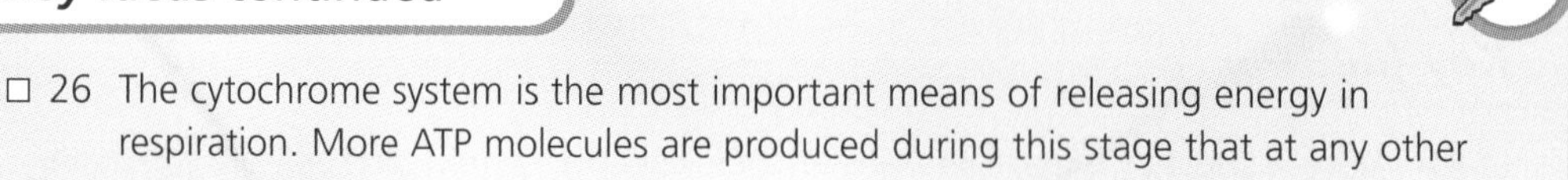

- ☐ 26 The cytochrome system is the most important means of releasing energy in respiration. More ATP molecules are produced during this stage that at any other stage.
- ☐ 27 Mitochondria are sausage-shaped with inner membranes folded to form cristae enclosing a central matrix.
- ☐ 28 The anaerobic phase of respiration is glycolysis which does not require oxygen.
- ☐ 29 Glycolysis produces very little ATP.
- ☐ 30 In anaerobic conditions in animal tissue, the pyruvic acid produced by glycolysis is converted to lactic acid as a final metabolic product.
- ☐ 31 In anaerobic conditions in plant tissue, the pyruvic acid is converted to carbon dioxide and ethanol as final metabolic products.

Topic Notes

The idea of respiration in cells is simple enough! It's the way that cells release energy trapped in their food. The food that's respired is usually glucose so it is the energy release from glucose that we study.

Top Tip

Foods that are respired are called respiratory substrates.

Glucose molecules are a source of chemical energy. Cells can't use this energy directly so if glucose solution is added to relaxed muscle fibres they will not contract.

The role and production of ATP

During respiration the energy in glucose is used to regenerate a substance called ATP. The ATP can be used to transfer chemical energy to cellular processes. When energy is released from glucose, ATP is regenerated from adenosine diphosphate (ADP) and inorganic phosphate (Pi), trapping the released energy. If ATP solution is added to relaxed muscle fibre they **will** contract instantly! On release of energy from ATP, the ADP and Pi are reformed.

ATP is important because it is a means of transferring chemical energy from food molecules such as glucose to energy-requiring processes such as contraction of muscle fibres. The diagram shows this idea.

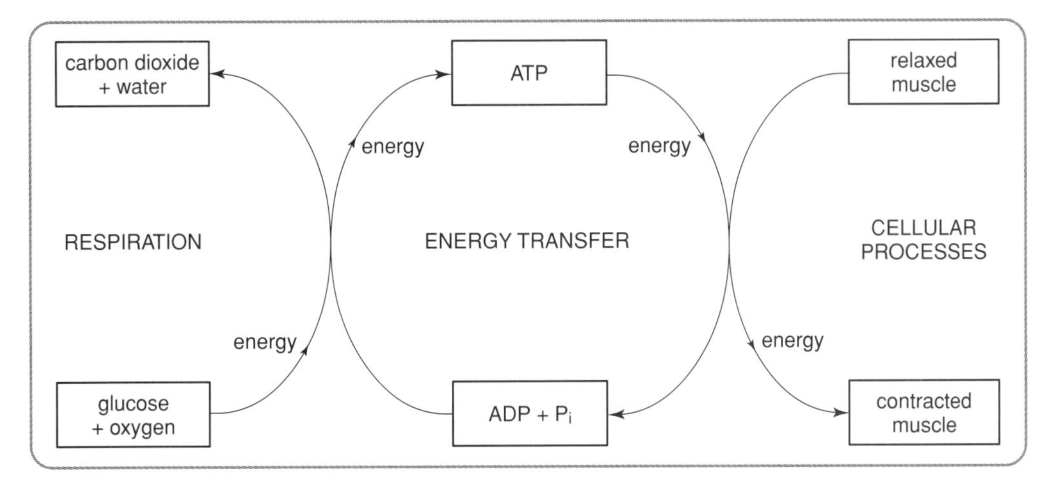

Figure 1.18 Transfer of chemical energy by ATP

Mitochondrion structure

Respiration occurs in the cytoplasm and the mitochondria of cells.

A mitochondrion is a sausage-shaped structure with an outer membrane and an inner membrane with folds called cristae enclosing a central matrix.

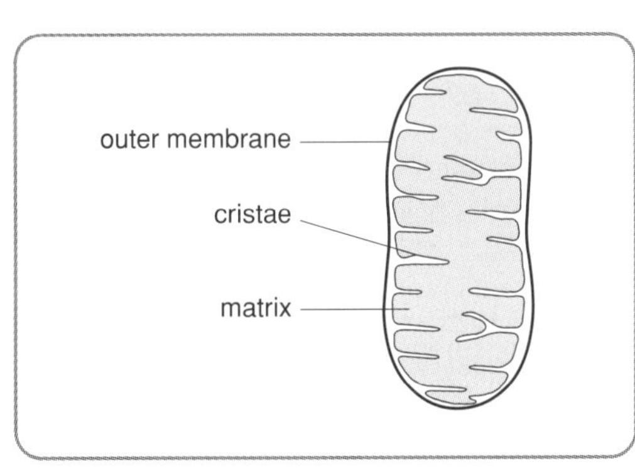

Figure 1.19 Features of a mitochondrion

Respiration is a series of reactions in which glucose is oxidised to form carbon dioxide. This process is accompanied by the synthesis of ATP from adenosine diphosphate (ADP) and inorganic phosphate (Pi). The oxidation of glucose happens in three stages and it is useful to study these one by one.
Oxidation is the loss of hydrogen from the respiratory substrate. The substance that gains the hydrogen is said to undergo reduction.

Top Tip

To help remember the meanings of oxidation and reduction, use the mnemonic **OILRIG**.

OIL **O**xidation **I**s **L**oss of hydrogen

RIG **R**eduction **I**s **G**ain of hydrogen

Learn respiration in three stages!

- Stage 1 – glycolysis
- Stage 2 – the Krebs Cycle
- Stage 3 – the cytochrome system

Top Tip

You will probably find that the features of each stage are easier to memorise as diagrams like those shown. Practise drawing the diagrams from memory!

Glycolysis

These are the main features of glycolysis you need to remember for the exam:

- Happens in the cytoplasm.
- Does not require oxygen.
- Converts the 6-carbon (6C) sugar glucose to two molecules of 3-carbon (3C) pyruvic acid.
- Provides a net gain of two molecules of ATP.
- Releases hydrogen.
- Hydrogen is carried away by NAD as NADH.

Top Tip

Two molecules of ATP are used to start the breakdown of a glucose molecule but four molecules of ATP are produced. This gives a **net** gain of two!

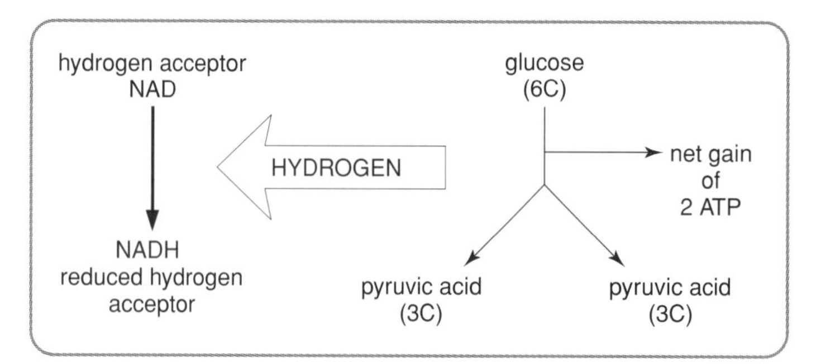

Figure 1.20 The main features of glycolysis

The Krebs cycle

These are the main features of the Krebs cycle you need to remember for your exam:

- Happens in the matrix of the mitochondria.
- It is aerobic. This means it requires oxygen.
- Starts with a 2-carbon (2C) acetyl group derived from pyruvic acid produced in glycolysis.
- The acetyl group joins with co-enzyme A (CoA) to form acetyl-CoA.
- Acetyl-CoA reacts with a 4-carbon (4C) compound to form a 6-carbon compound called citric acid.
- Citric acid is gradually converted back to the 4C compound. Its carbons are lost in carbon dioxide.
- Hydrogen is released and carried away by NAD as NADH.

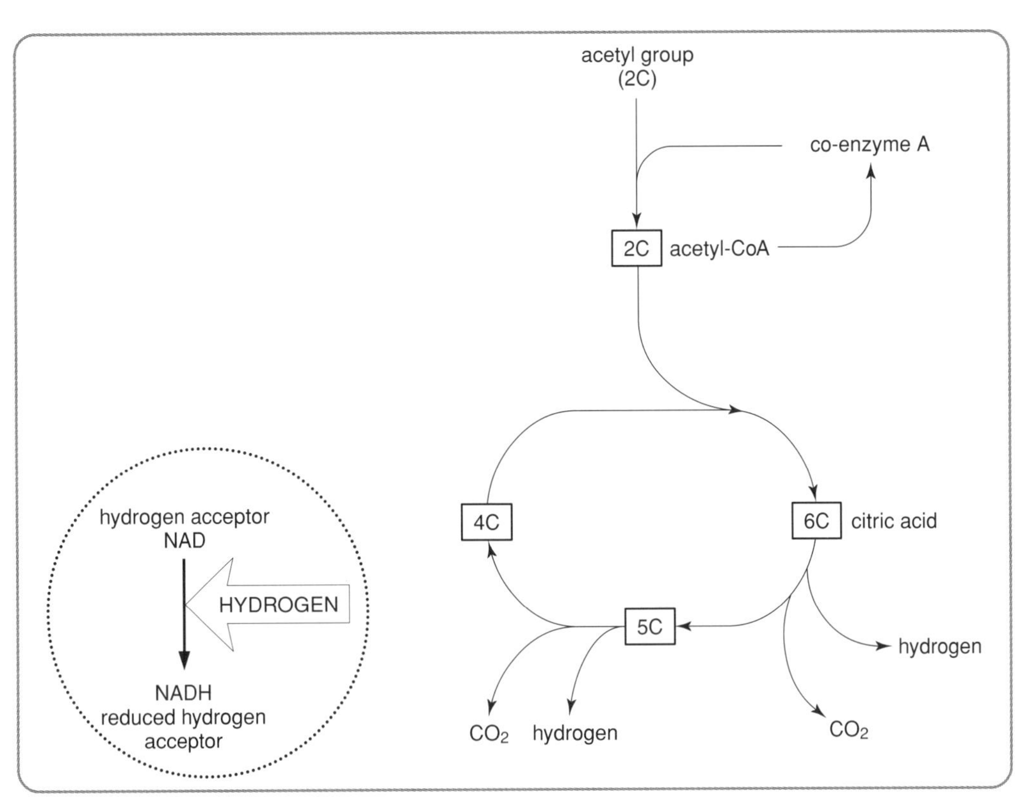

Figure 1.21 The main features of the Krebs cycle

The cytochrome system

These are the main feature of the cytochrome system you need to remember for your exam:

- It occurs on the cristae of the mitochondria.
- It is aerobic.
- Hydrogen is passed through a series of carriers and finally combined with oxygen to from water.
- Oxygen is the final hydrogen acceptor.
- Most ATP is produced at this stage.

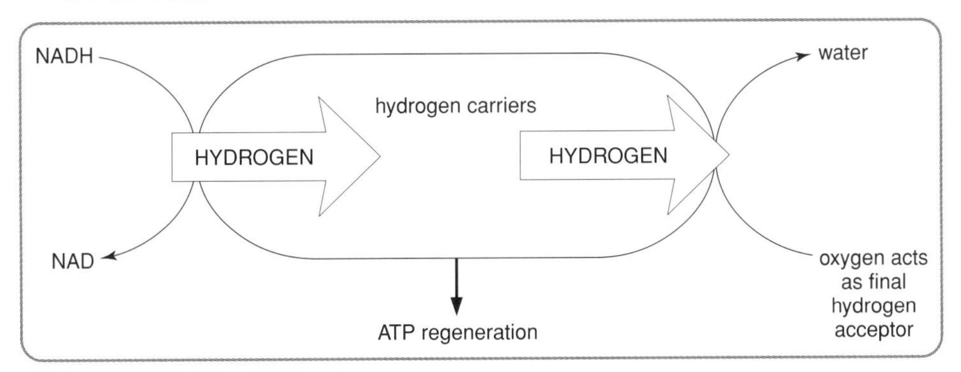

Figure 1.22 The main features of the cytochrome system

Glucose is the commonest respiratory substrate as it is the final product of the digestion of carbohydrates in the diet. Fats and proteins in the diet are alternative respiratory substrates.

Distinction between aerobic and anaerobic phases of respiration

Glycolysis is the anaerobic phase of respiration. It produces only two molecules of ATP and its final metabolic product is pyruvic acid. The Krebs cycle and the cytochrome system form the aerobic phases of respiration. Thirty-six molecules of ATP are produced in the cytochrome system and the final metabolic products are carbon dioxide from Krebs cycle and water from the cytochrome system.

Cells can respire anaerobically involving glycolysis only. In animal cells the pyruvic acid is changed into lactic acid. In plants the pyruvic acid is converted into carbon dioxide and ethanol. Anaerobic respiration produces only two molecules of ATP.

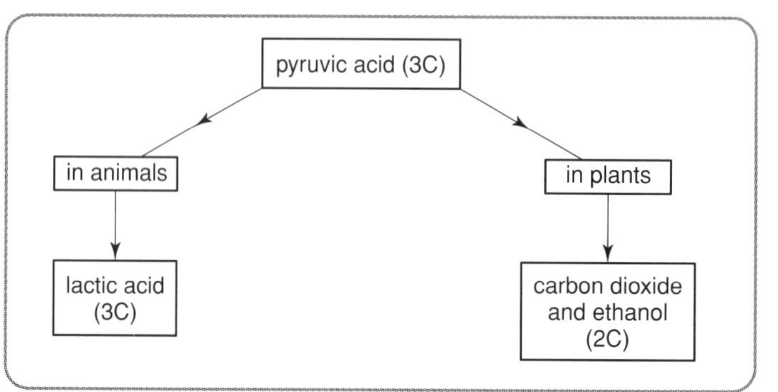

Figure 1.23 The main features of aerobic respiration

Exam Questions

marks

SBQ 1 The diagram below represents stages in aerobic respiration in mammal muscle cells.

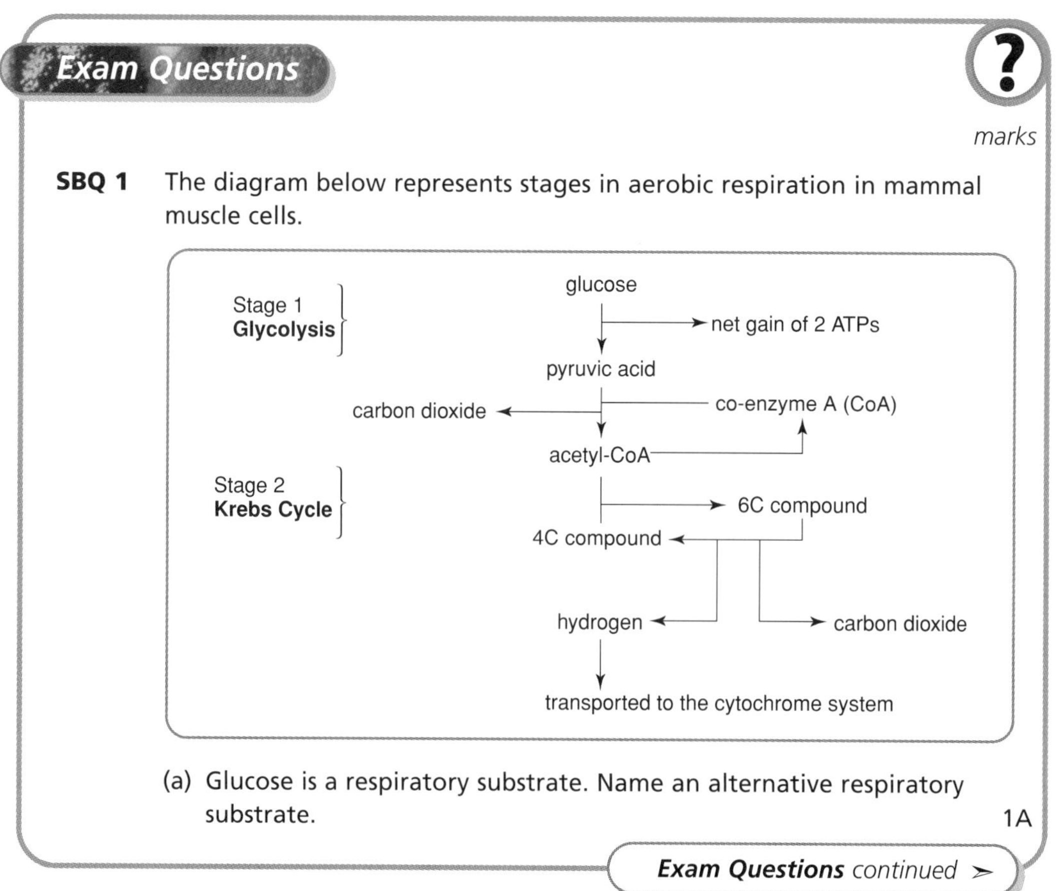

(a) Glucose is a respiratory substrate. Name an alternative respiratory substrate. 1A

***Exam Questions** continued* ➢

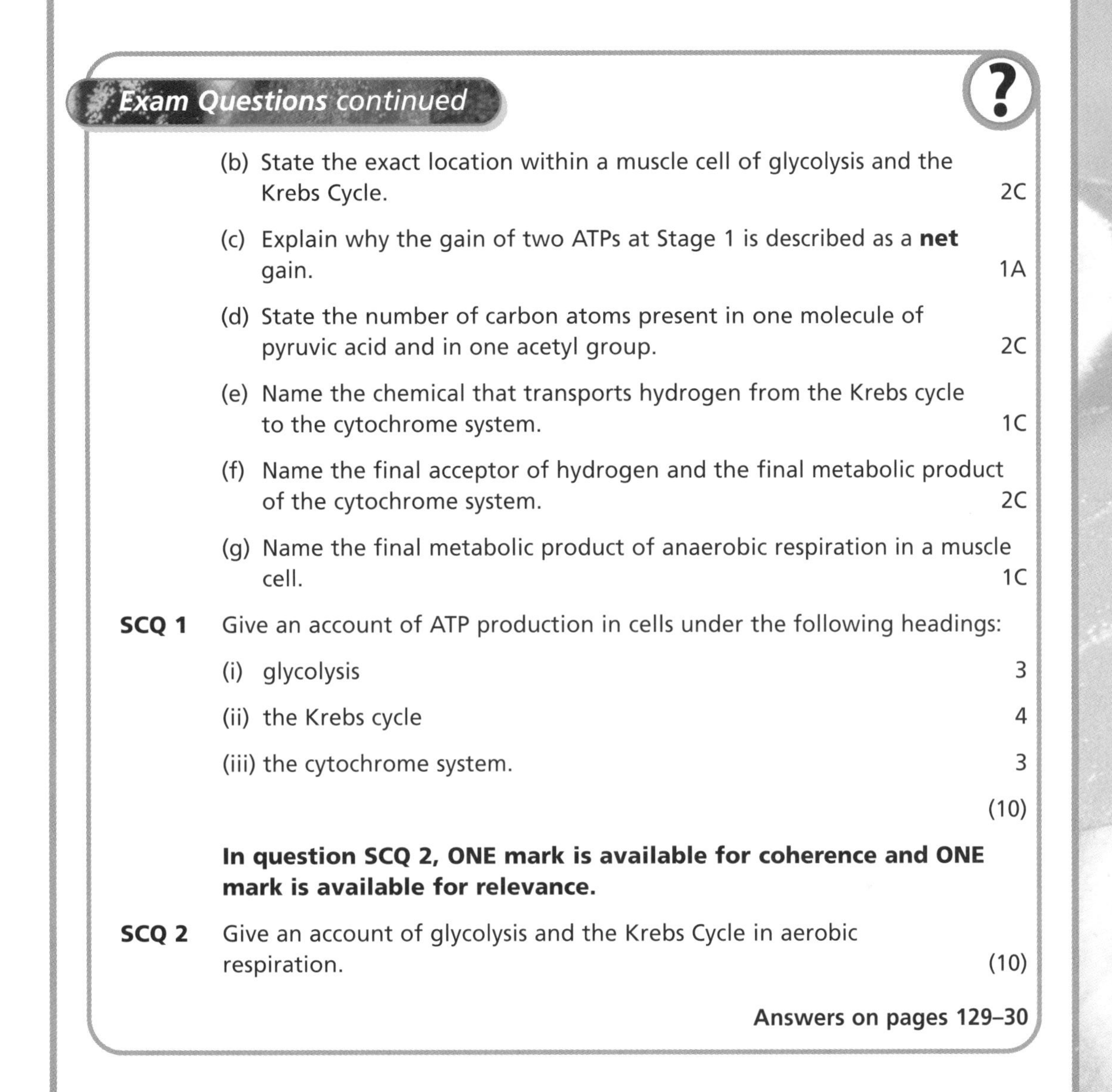

Exam Questions continued

(b) State the exact location within a muscle cell of glycolysis and the Krebs Cycle. 2C

(c) Explain why the gain of two ATPs at Stage 1 is described as a **net** gain. 1A

(d) State the number of carbon atoms present in one molecule of pyruvic acid and in one acetyl group. 2C

(e) Name the chemical that transports hydrogen from the Krebs cycle to the cytochrome system. 1C

(f) Name the final acceptor of hydrogen and the final metabolic product of the cytochrome system. 2C

(g) Name the final metabolic product of anaerobic respiration in a muscle cell. 1C

SCQ 1 Give an account of ATP production in cells under the following headings:

(i) glycolysis 3

(ii) the Krebs cycle 4

(iii) the cytochrome system. 3

(10)

In question SCQ 2, ONE mark is available for coherence and ONE mark is available for relevance.

SCQ 2 Give an account of glycolysis and the Krebs Cycle in aerobic respiration. (10)

Answers on pages 129–30

1.4 Synthesis and Release of Proteins

Key Ideas

- ☐ 1 There is a huge variety of proteins in living things.
- ☐ 2 Proteins contain the elements carbon, hydrogen, oxygen and nitrogen.
- ☐ 3 Proteins are built up from sub-units called amino acids.
- ☐ 4 There are about 20 different amino acids that occur naturally.
- ☐ 5 Amino acids are linked together by peptide bonds to form polypeptide chains.
- ☐ 6 The order or sequence of amino acids is important in the structure and function of proteins.

***Key Ideas** continued* ➢

Key Ideas continued

- ☐ 7 Proteins can be classified as either fibrous (e.g. collagen) or globular (e.g. enzymes, some membrane proteins, some hormones and antibodies).
- ☐ 8 An organism's genetic information consists of coded instructions for making protein and is located in the chromosomes usually found in the nuclei of its cells.
- ☐ 9 The chromosomes are chains of genes.
- ☐ 10 Genes are regions of chromosomal DNA.
- ☐ 11 DNA is a complex substance that carries the genetic information of the organism as a code.
- ☐ 12 A DNA molecule is in the form of a double helix. This means that it is made up of two strands twisted into a spiral shape.
- ☐ 13 Each strand of DNA is made up of a chain of units called nucleotides.
- ☐ 14 A DNA nucleotide unit has three parts, a deoxyribose sugar, a phosphate group and a nitrogenous base.
- ☐ 15 There are four types of DNA nucleotide that contain different bases. The different bases are adenine, guanine, thymine and cytosine.
- ☐ 16 The nucleotides of each strand are connected together through their sugars and phosphates to give a strong sugar-phosphate backbone.
- ☐ 17 The two strands are connected by weak hydrogen bonds joining the bases.
- ☐ 18 The bases are connected in a specific way with adenine only pairing with thymine, and guanine only pairing with cytosine. This is called complementary base pairing.
- ☐ 19 It is the sequence of bases along one of the strands that makes up the genetic code, specifying the types of protein that can be made.
- ☐ 20 DNA has a triplet code. Each group of three bases along a strand codes for a specific amino acid.
- ☐ 21 The sequence of base triplets on the DNA strand determines the sequence of amino acids found in the protein.
- ☐ 22 Before a cell can divide, its DNA must replicate. This process copies the genetic information coded in the cell's DNA.
- ☐ 23 Replication ensures that daughter cells can receive an exact copy of the mother cell's genetic information.
- ☐ 24 Replication requires: the original DNA molecule to act as a template, enzymes to control the process, a supply of free DNA nucleotides and ATP for energy.
- ☐ 25 The DNA untwists and the hydrogen bonds joining the two strands of the original molecule break allowing the strands to pull apart or unzip.
- ☐ 26 Free DNA nucleotides align themselves beside their complementary base pair partners on the exposed DNA strand and link with them by hydrogen bonds.
- ☐ 27 The free nucleotides then join up together through sugar phosphate backbones to form the new strands of DNA.
- ☐ 28 Each original strand then twists into a new double helix with a newly formed strand. This produces two identical copies of the original DNA.
- ☐ 29 After complete replication of a cell's DNA, each chromosome consists of two parts called chromatids. Each chromatid has a copy of the DNA of the original chromosome.
- ☐ 30 Cells contain another type of nucleic acid called RNA.

Key Ideas continued ➢

Key Ideas continued

- ☐ 31 Two forms of RNA, called messenger RNA (mRNA) and transfer RNA (tRNA), are important in the process of protein synthesis in cells.
- ☐ 32 RNA differs in several ways from DNA. It is single-stranded and its nucleotides have ribose sugar instead of deoxyribose and the base uracil replaces thymine.
- ☐ 33 mRNA synthesis requires: the DNA to act as a template, a supply of free RNA nucleotides, enzymes to control the process and ATP for energy.
- ☐ 34 When a protein requires to be synthesised, the appropriate DNA in the nucleus of a cell untwists and unzips by the breaking of hydrogen bonds between its base pairs.
- ☐ 35 Free nucleotides of RNA align themselves beside the exposed bases of the DNA code and link to the DNA bases by hydrogen bonds.
- ☐ 36 The RNA nucleotides that contain uracil bases pair with the adenine bases on the DNA strand.
- ☐ 37 The nucleotides of the RNA join up through sugar phosphate backbones to form a new molecule called messenger RNA. A group of three bases on mRNA is called a codon.
- ☐ 38 The messenger RNA contains a complementary copy of the DNA genetic code for a specific protein.
- ☐ 39 The production of mRNA from a DNA template is called transcription.
- ☐ 40 Proteins are synthesised at ribosomes, and mRNA molecules leave the nucleus via pores and attach themselves to vacant ribosomes.
- ☐ 41 Molecules of transfer RNA, each carrying a group of three bases called an anticodon, pick up specific amino acids from the cytoplasm of the cell.
- ☐ 42 There are 20 different tRNA molecules, one type for each of the 20 different amino acids.
- ☐ 43 The tRNA molecules carrying a specific amino acid align themselves against the mRNA molecule at a ribosome. The anticodons form hydrogen bonds with the complementary codons of mRNA.
- ☐ 44 This process brings the amino acids carried by the tRNA into line and they link up by peptide bonds that form between them to produce polypeptides.
- ☐ 45 The order of the amino acids in the polypeptide has been determined by the sequence of bases on the original DNA.
- ☐ 46 The process of formation of the polypeptides is called translation.
- ☐ 47 After translation is complete, the protein formed is passed through the channels of the endoplasmic reticulum.
- ☐ 48 The protein is passed to the Golgi apparatus where it is processed and packaged into vesicles ready for secretion.
- ☐ 49 Secretory vesicles fuse with the plasma membrane during secretion and the proteins are released from the cell.

Topic Notes

The functional variety of proteins

Proteins form the structure of living organisms and are vital in the functioning of their cells. For this reason there is a huge variety of different proteins.

Proteins are made up from chains of units called amino acids. There are 20 different amino acids found naturally and many proteins contain thousands of amino acids linked together. The structure and function of a protein is dependent on the amino acid sequence it has.

The amino acids are linked by peptide bonds and the chain is folded to form the protein shape.

In **fibrous** proteins, the chains are folded into parallel strands and cross-linked for strength. These types of protein are used to make structural materials, such as collagen, important in skin and bone.

In **globular** proteins, the chains are folded into ball shapes that are held together with weak bonds giving a delicate structure. These types of protein have very specific shapes and form enzymes, membrane proteins, some hormones and antibodies.

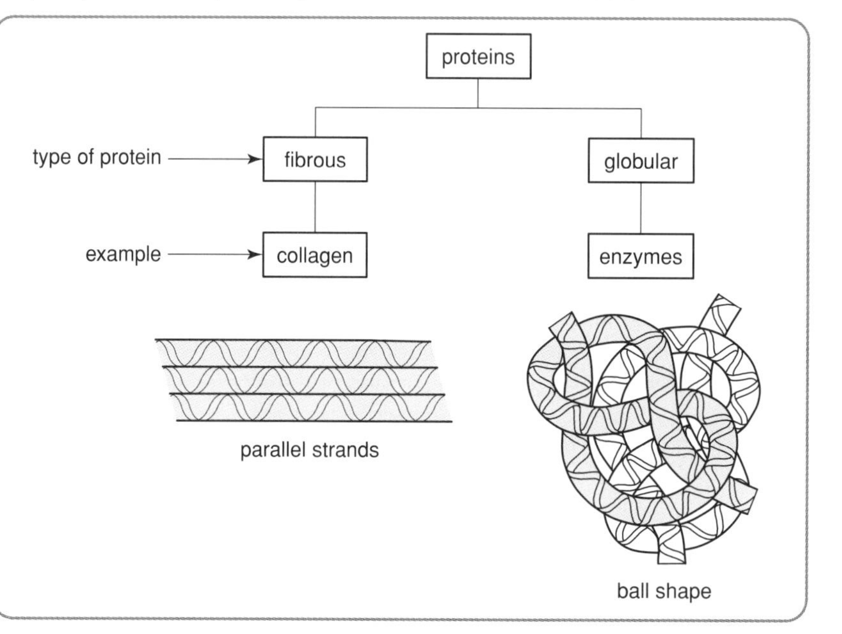

Figure 1.24 Protein types

The proteins that a cell makes are determined by its genetic information. The genetic information is encoded information for the synthesis of proteins.

A cell has its genetic information packaged as chromosomes within its nucleus. A chromosome is a string of genes each of which is a region of chromosomal DNA. Each gene carries a code for the synthesis of a specific protein.

Top Tip

Cells .... contain nuclei.
Nuclei .... contain chromosomes.
Chromosomes .... contain genes.
Genes .... contain DNA.
DNA .... contains genetic information.

Each piece of genetic information is an instruction for making a specific protein. The instruction is coded into the structure of DNA.

The genes on a chromosome are lengths of DNA that code for a particular protein. The genetic code determines the order of amino acids that determines the type of protein and its function.

DNA structure

DNA is made up of two strands of nucleotides twisted into a spiral called a double helix.

Each nucleotide is made up of a phosphate, a deoxyribose sugar and a base as shown in the diagram.

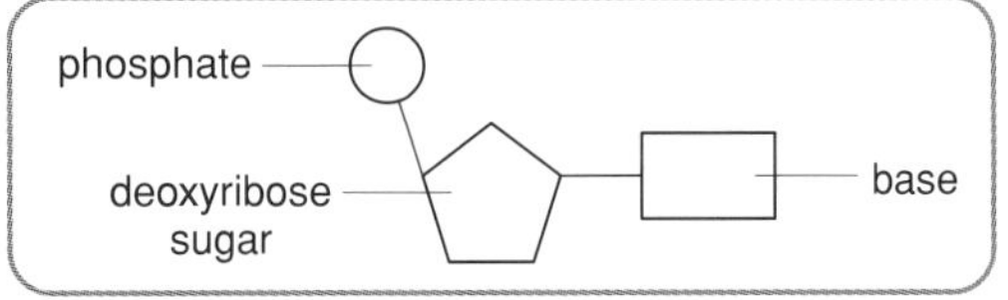

Figure 1.25 Nucleotide structure

There are four different bases, adenine (A), thymine (T), cytosine (C) and guanine (G), and so four different DNA nucleotides.

Top Tip

In your exam, you must be able to **name** the bases **not just give their letters**.

The nucleotides in a DNA strand are joined together by strong sugar-phosphate bonds as shown in the diagram.

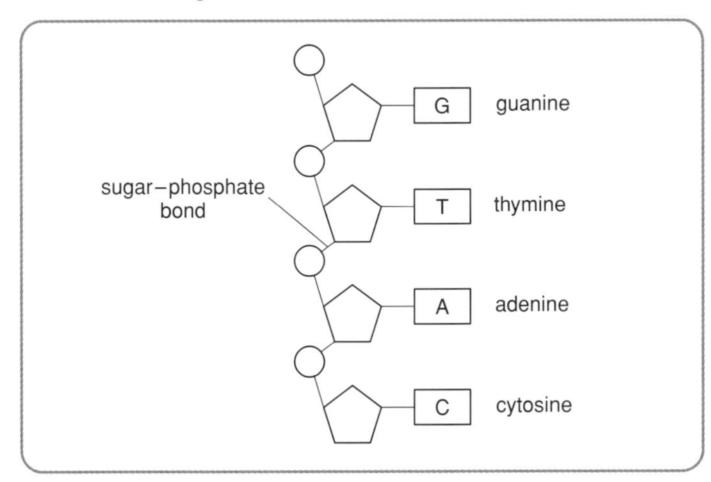

Figure 1.26 DNA strand

The two strands are held together by weak hydrogen bonds that form between the bases of each strand. The base pairing is specific, only certain bases pair together. This is called complementary base pairing.

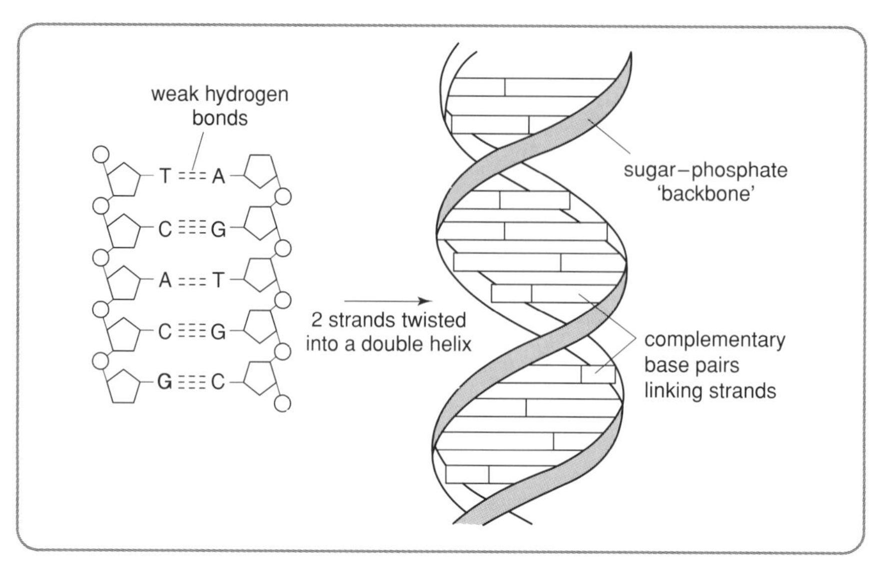

Figure 1.27 Complementary base pairing

Adenine pairs with thymine; cytosine pairs with guanine.

The order of the bases on DNA determines the sequence of amino acids in the protein it codes for.

Each group of three bases along a DNA strand is called a triplet. Each triplet codes for a specific amino acid as shown on the diagram.

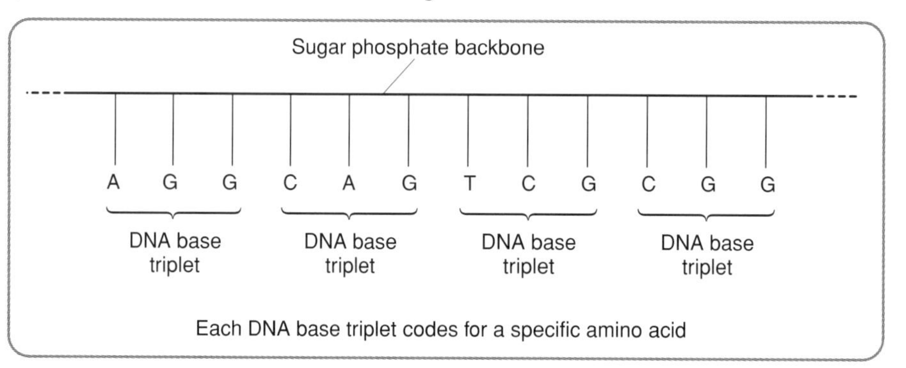

Figure 1.28 Part of one strand of a DNA molecule

DNA replicates before a cell divides by mitosis or meiosis. The result of DNA replication is the formation of two exact copies of the original DNA molecule. This is important as it ensures that each daughter cell produced carries an exact copy of each gene it requires to carry out all of its functions.

The main events in DNA replication are summarised below:

1 The DNA double helix untwists.
2 The weak hydrogen bonds break and the DNA unzips.
3 Free DNA nucleotides present in the cytoplasm enter the nucleus and join with their complementary base pair partners.

4 Sugar-phosphate bonds join the nucleotides together forming the sugar-phosphate 'backbones' of the new DNA strands.

5 The newly-formed DNA molecules now retwist and form two double helices.

After the entire DNA on a chromosome has replicated, the chromosomes appear composed of two chromatids. This general idea is shown on the diagram.

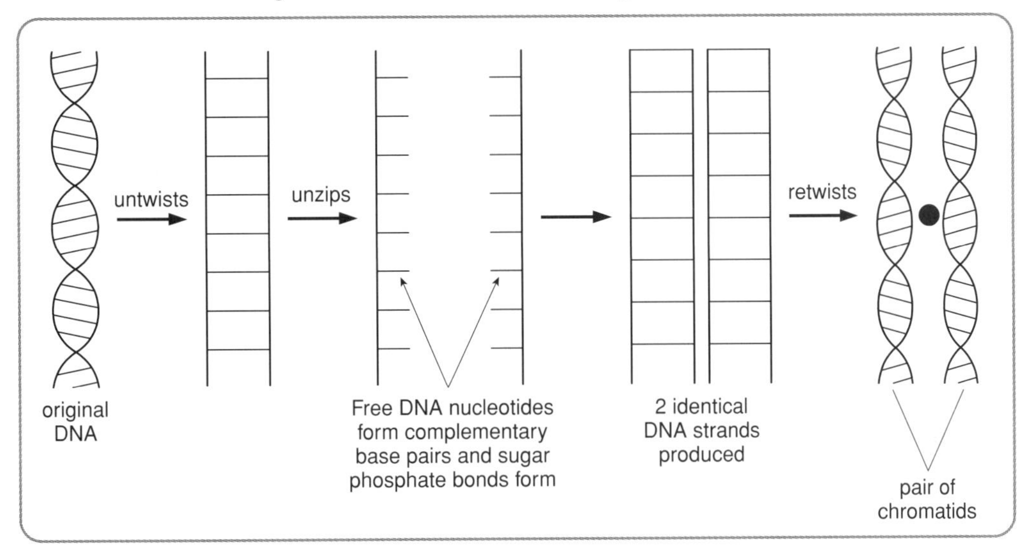

Figure 1.29 DNA replication

RNA

RNA is a second type of nucleic acid. Like DNA it is made up of nucleotides.

The table shows the differences between RNA and DNA.

Feature	DNA	RNA
Number of strands	two	one
Sugar	deoxyribose	ribose
Bases	adenine, **thymine**, cytosine and guanine	adenine, **uracil**, cytosine and guanine

Top Tip

In your exam, you may be asked to state the differences between DNA and RNA – make sure that you compare them fully as shown in the table.

RNA exists in two different forms: messenger RNA (mRNA) and transfer RNA (tRNA).

mRNA is formed in the nucleus of the cell using DNA as a template. This process is called transcription.

Transcription occurs in the following steps:

1 The DNA template untwists and unzips.
2 Free RNA nucleotides enter the nucleus and form complementary base pairs with one strand of the DNA
3 Weak hydrogen bonds form between complementary base pairs. Uracil pairs with adenine.
4 Sugar-phosphate bonds form between the RNA nucleotides to form mRNA.

The mRNA strand breaks away from the DNA strand and passes through a pore of the nucleus into the cytoplasm where it attaches to a ribosome.

Each group of three bases on the mRNA strand is called a codon. Each codon codes for an amino acid.

tRNA is found in the cytoplasm. Each group of three bases on tRNA is called an anticodon.

Each tRNA picks up and carries a specific amino acid to the ribosome to be joined into a forming protein. The amino acid picked up is specific to the anticodon on the tRNA.

Once at the ribosome each tRNA anticodon aligns with its complementary codon on mRNA bringing each amino acid into sequence.

Peptide bonds form between the adjacent amino acids forming a protein molecule.

The formation of the protein from the mRNA at a ribosome is called translation.

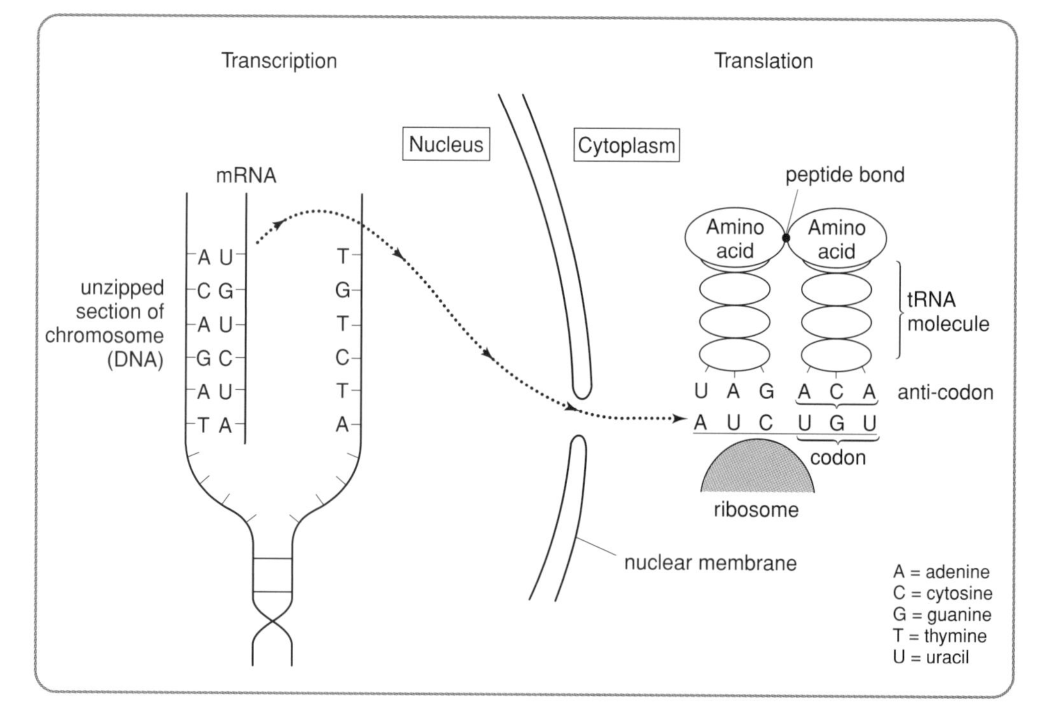

Figure 1.30 Transcription and translation

The completed protein may be transported through the channels of the rough endoplasmic reticulum (ER) and passed to the Golgi apparatus. The Golgi apparatus processes and packages the protein ready for secretion.

Exam Questions

Marks

SBQ 1 The diagram below represents part of a DNA molecule during replication.

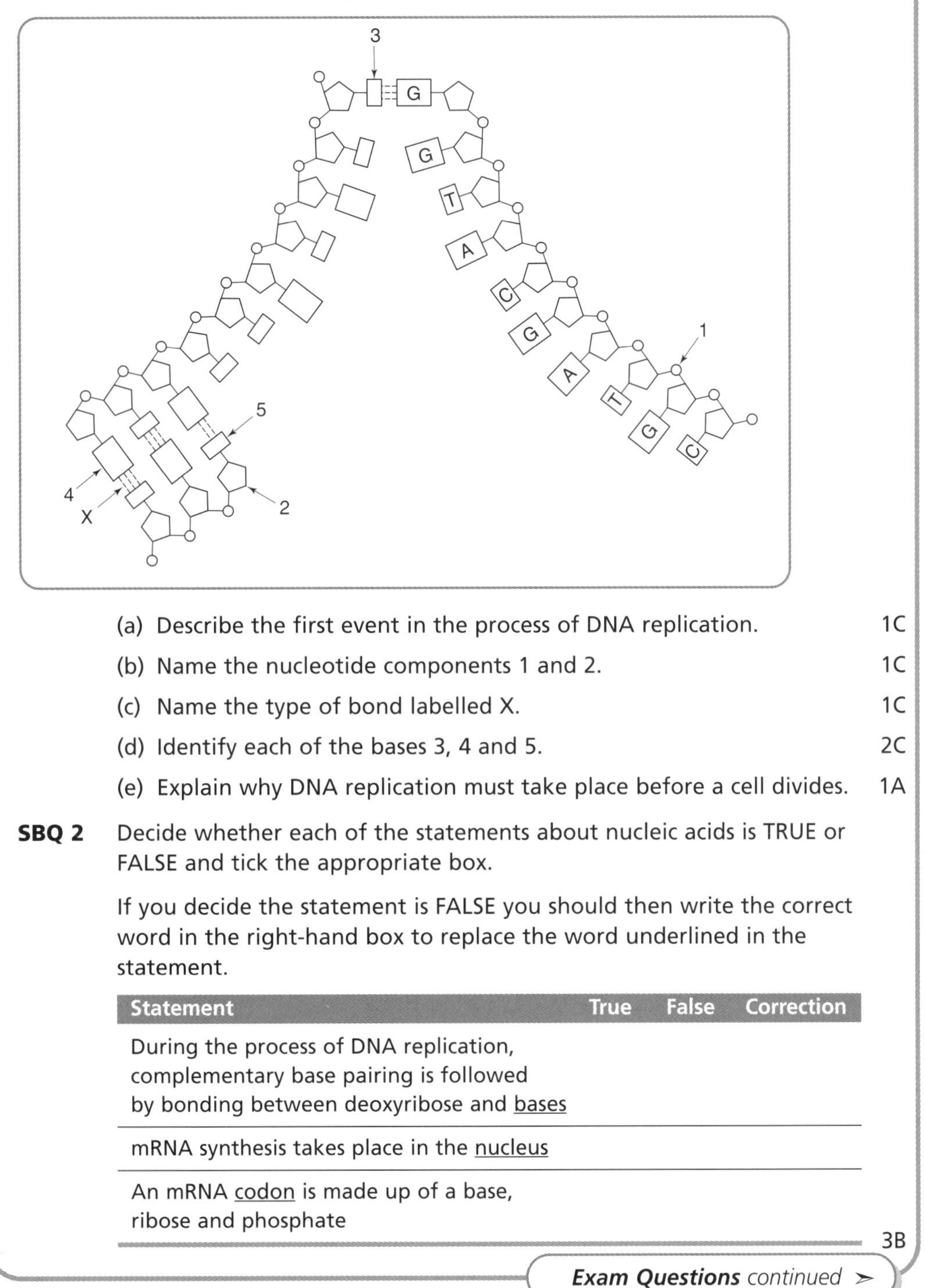

(a) Describe the first event in the process of DNA replication. 1C

(b) Name the nucleotide components 1 and 2. 1C

(c) Name the type of bond labelled X. 1C

(d) Identify each of the bases 3, 4 and 5. 2C

(e) Explain why DNA replication must take place before a cell divides. 1A

SBQ 2 Decide whether each of the statements about nucleic acids is TRUE or FALSE and tick the appropriate box.

If you decide the statement is FALSE you should then write the correct word in the right-hand box to replace the word underlined in the statement.

Statement	True	False	Correction
During the process of DNA replication, complementary base pairing is followed by bonding between deoxyribose and <u>bases</u>			
mRNA synthesis takes place in the <u>nucleus</u>			
An mRNA <u>codon</u> is made up of a base, ribose and phosphate			

3B

Exam Questions *continued* ➢

Exam Questions continued

SBQ 3 In a molecule of DNA, the base sequence AGC codes for the amino acid serine. Using the initial letters of the bases, write the base sequence of the tRNA anticodon to which serine becomes attached. 1B

SBQ 4 (a) Describe the roles of tRNA in protein synthesis. 2B/A

(b) Describe the function of the rough endoplasmic reticulum (ER) and the Golgi apparatus in a cell. 2C

SBQ 5 (a) Complete the diagram below which shows information about protein classification. 1C/B

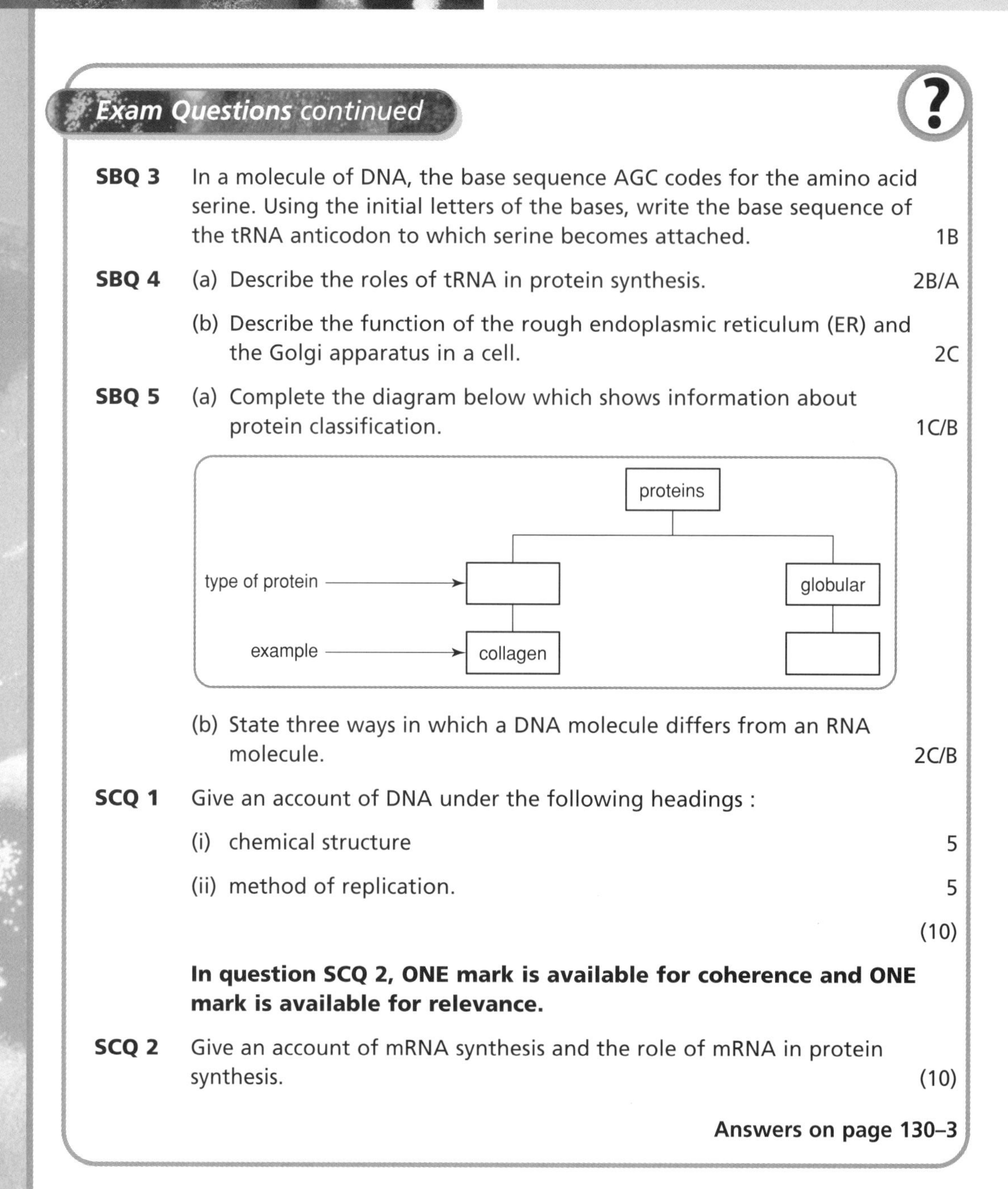

(b) State three ways in which a DNA molecule differs from an RNA molecule. 2C/B

SCQ 1 Give an account of DNA under the following headings :

(i) chemical structure 5

(ii) method of replication. 5

(10)

In question SCQ 2, ONE mark is available for coherence and ONE mark is available for relevance.

SCQ 2 Give an account of mRNA synthesis and the role of mRNA in protein synthesis. (10)

Answers on page 130–3

1.5 Cellular Response in Defence

Key Ideas

- ☐ 1 Viruses are extremely small structures that contain a nucleic acid, either DNA or RNA, surrounded by a protein coat.
- ☐ 2 Viruses invade specific host cells.
- ☐ 3 Viruses can only reproduce or replicate inside a host cell. They alter the host cell's instructions and its metabolism to allow viral nucleic acid to replicate resulting in the release of large numbers of viruses.
- ☐ 4 The virus attaches to the surface of the host cell. Then the viral nucleic acid enters the cell and takes over control of the cell altering its metabolism. The viral nucleic acid is copied or replicated using the host cell's nucleotides and ATP. The viral nucleic acid then instructs the host cell to produce protein coats using the host cell's amino acids and new viruses are assembled.
- ☐ 5 Phagocytosis and antibody production are the main cellular defence mechanisms in animals.
- ☐ 6 Phagocytosis is carried out by a group of white blood cells called phagocytes.
- ☐ 7 Phagocytes move towards foreign particles such as bacteria. They engulf the bacteria enclosing them in a vacuole. Lysosomes move towards the vacuole fusing with it and releasing enzymes into the vacuole to digest the bacteria.
- ☐ 8 Another group of white blood cells called lymphocytes recognise invading foreign organisms and produce proteins called antibodies in response to the presence of foreign antigens on their surface.
- ☐ 9 An antigen is an organic substance such as protein or carbohydrate that is recognised as being foreign by the lymphocytes.
- ☐ 10 Lymphocytes produce specific antibodies in response to a specific antigen.
- ☐ 11 Antibodies cause destruction of antigens.
- ☐ 12 The presence of antigens on tissues used for transplantation causes production of antibodies that can lead to rejection of the transplant. Immunosuppressor drugs can be used to reduce this problem during and after transplants.
- ☐ 13 Plants protect themselves by producing a variety of toxic compounds or substances that act as barriers to isolate an injured area.
- ☐ 14 The toxic compounds are tannins, cyanide and nicotine.
- ☐ 15 Many plants produce tannins that give them an unpleasant taste and so discourage damage caused by grazing herbivores.
- ☐ 16 Cyanide and nicotine are toxic chemicals produced by some plants in response to damage to their leaves.
- ☐ 17 Resin is a sticky barrier substance produced by some plants when they have been damaged or injured. It prevents the spread of infection to other parts of the plant.

Topic Notes

The nature of viruses

Viruses are extremely small structures that contain a nucleic acid, either DNA or RNA that is surrounded by a protein coat.

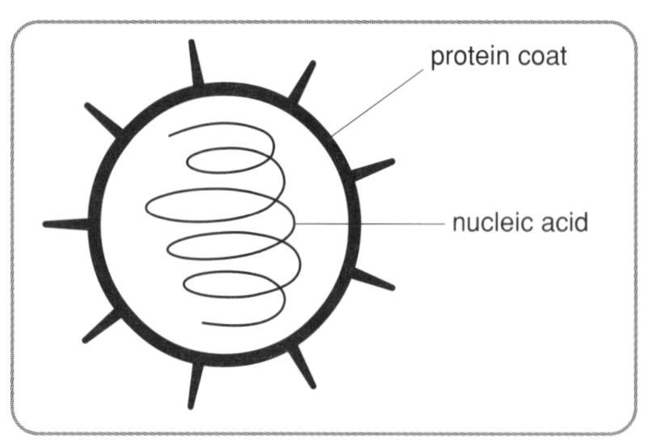

Figure 1.31 The main features of a virus

They are pathogens, which means that they cause disease. They attack or infect specific host cells. They can only reproduce or replicate inside a host cell. The viruses alter the host cell's metabolism to allow them to replicate resulting in the release of large numbers of new viruses.

Viral replication

The stages in viral replication are shown in Figure 1.32 and listed below:

1. The virus attaches to the surface of the host cell. →
2. The viral nucleic acid (DNA/RNA) enters the cell. →
3. The viral nucleic acid takes over control of the cell and alters the cell's metabolism. It stops the cell's normal DNA/RNA replication and protein synthesis. →
4. The viral nucleic acid is copied or replicated using the host cell's nucleotides and ATP. →
5. Each new copy of the viral nucleic acid then instructs the host cell to produce protein coats using the host cell's amino acids and ATP. The host cell's ribosomes are the organelles at which the viral protein coats are synthesised. →
6. The virus particles are assembled when the protein coats each surround a viral nucleic acid molecule. →
7. Large numbers of viruses are released when the cell bursts open. This is called lysis.

Top Tip

Learning the virus flowchart could be really useful for answering an extended response question on this topic.

Cellular defence mechanisms in animals

Phagocytosis and antibody production are the main cellular defence mechanisms in animals.

Phagocytosis

Phagocytosis is carried out by a group of white blood cells called phagocytes. Phagocytes have many lysosomes that contain powerful protein-digesting enzymes. The phagocytes

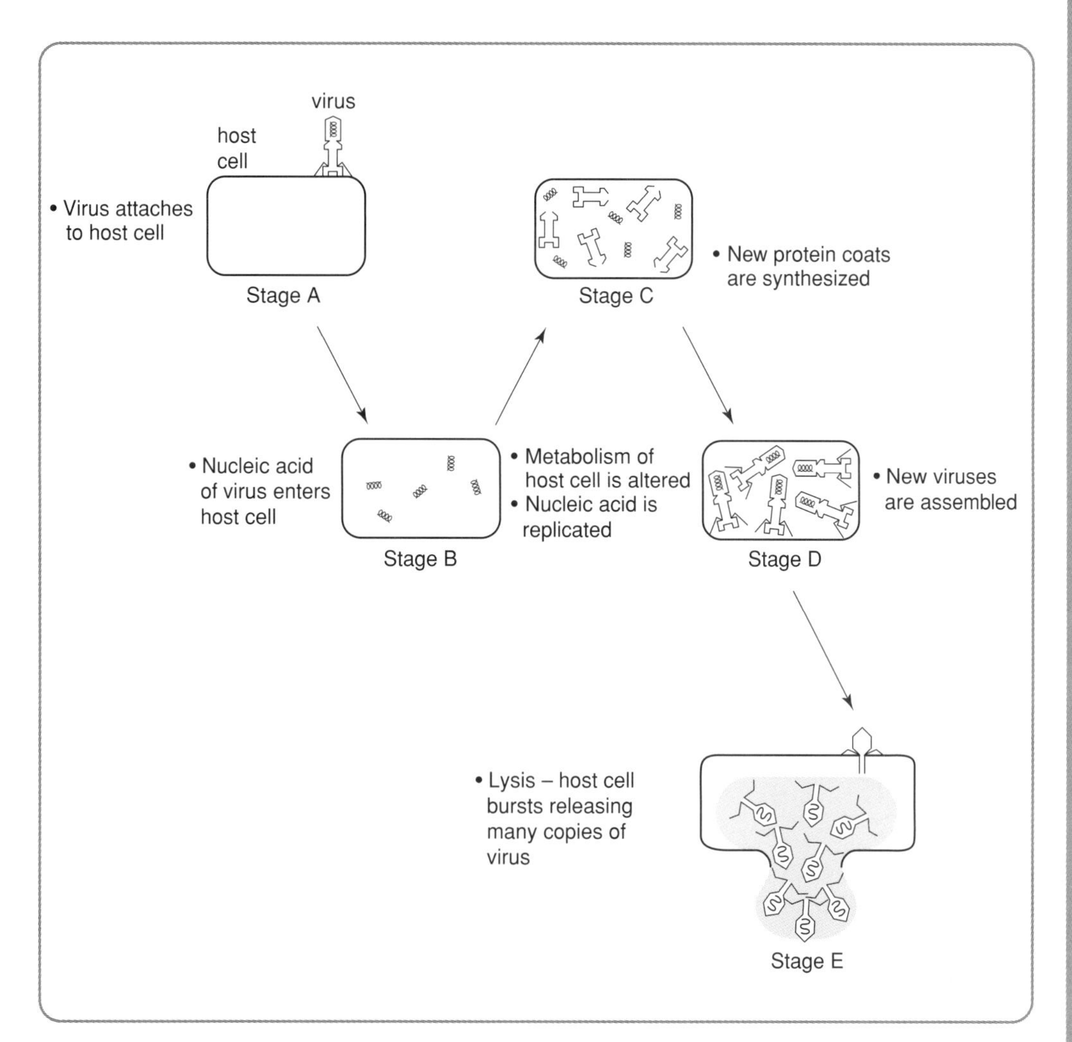

Figure 1.32 The stages in viral replication

move towards foreign particles such as bacteria. They surround and engulf a bacterium enclosing it in a vacuole. The lysosomes move towards the vacuole fusing with the vacuole membrane and releasing the enzymes into the vacuole to digest the bacterium.

Top Tip

Make sure that you know the role of the lysosomes in phagocytosis.

Antibody Production

Another group of white blood cells called lymphocytes recognise invading foreign material and produce Y-shaped proteins called antibodies in response to the presence of foreign antigens on the surface of the invader.

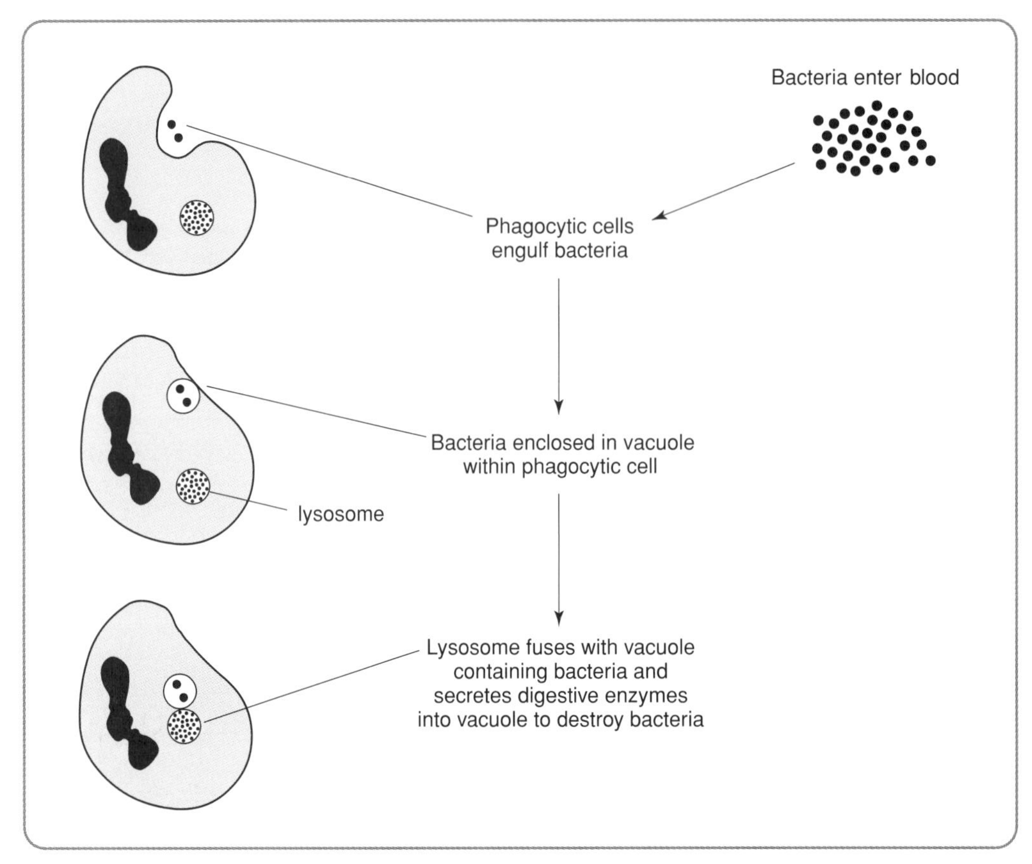

Figure 1.33 The stages of phagocytosis

Top Tip

The **antigens** cause **anti**bodies to be **gen**erated by lymphocytes.

An antigen is an organic substance such as protein or carbohydrate that is recognised as being foreign by the lymphocytes. Once a lymphocyte has recognised an antigen, it replicates so that it can mass-produce the required antibodies. The lymphocytes produce **specific** antibodies in response to the antigen. The antibodies cause the destruction of the antigen.

Top Tip

In some of the more difficult questions on this topic you must know what is meant by the **specificity** of antibodies.

If a living organ such as a heart or kidney is transplanted from one person to another, there is a risk of rejection because the recipient's lymphocytes reacts against the foreign antigens on the transplanted organ.

Cellular defence mechanisms in plants

Plants can protect themselves by producing a variety of toxic compounds or by the production of a substance that acts as a barrier and isolates the injured area.

The toxic compounds that you have to know are tannins, cyanide and nicotine.

Toxic compound	Explanation for effect
Tannins	Tannins give plants an unpleasant taste and discourage grazing by herbivores.
Nicotine	Nicotine is a toxic compound produced to discourage grazing by herbivores.
Cyanide	Cyanide is an enzyme inhibitor that stops aerobic respiration. Plants that produce cyanide when damaged are cyanogenic.

Top Tip

It is useful to know the term 'Cyanogenic'.

Top Tip

Three **N**asty **C**hemicals = **T**annins, **N**icotine, and **C**yanide.

The only barrier substance that you need to know is **resin**. Many plants produce resin when they have been damaged or injured. Resin is a sticky substance produced by the damaged cells that prevents the spread of infection to other parts of the plant.

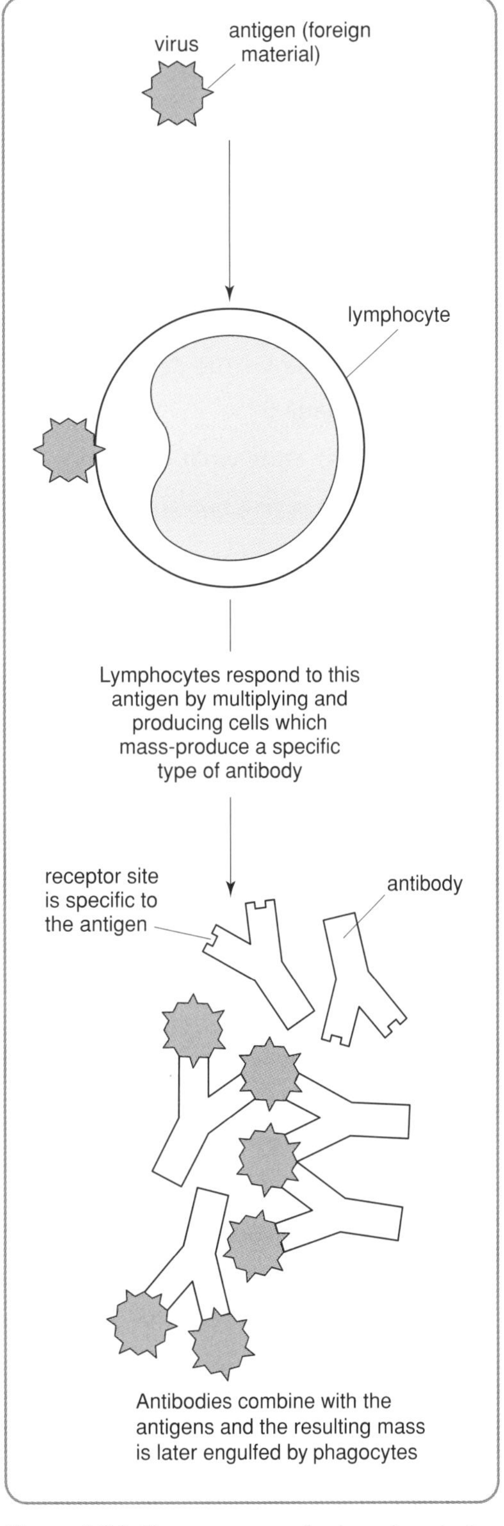

Figure 1.34 The response of a lymphocyte to a foreign antigen

Exam Questions

Marks

SBQ 1 The stages shown below occur during the invasion of a cell by a virus:

Stage 1 The viral nucleic acid enters the host cell.
↓
Stage 2 The virus alters the metabolism of the host cell.
↓
Stage 3
↓
Stage 4 The viral coat is synthesised.
↓
Stage 5 The new viruses are assembled.
↓
Stage 6

(a) State what occurs during Stages 3 and 6. 2C/B

(b) Name two substances that are supplied by the host cell in stage 4. 2C

(c) Name the host cell organelle which is the site for the synthesis of the viral coat at Stage 4. 1C

SBQ 2 During a viral infection, lymphocytes are stimulated to make chemicals that inactivate the viruses.

(a) Name the chemicals produced by the lymphocytes. 1C

(b) Which feature of viruses results in the production of these chemicals by the lymphocytes? 1C

SBQ 3 (a) Name the process by which some white blood cells surround, engulf and digest bacteria? 1C

(b) Name the organelles present in these white blood cells that contain powerful digesting enzymes. 1C

SBQ 4 (a) Name two toxic compounds produced by some plants to deter grazing by herbivores. 1C

(b) Name a chemical produced by some plants which acts as a barrier to prevent the spread of infection. 1C

SCQ 1 Write notes on each of the following:

(i) phagocytosis 4

(ii) antibody production 3

(iii) cellular defence mechanisms in plants. 3

(10)

In question SCQ 2, ONE mark is available for coherence and ONE mark is available for relevance.

SCQ 2 Give an account of the nature of viruses and their replication. (10)

Answers on page 134–5

Chapter 2

GENETICS AND ADAPTATION

Unit 2 is called 'Genetics and Adaptation'. It comprises the following topic areas:

2.1 Variation

2.2 Selection and Speciation

2.3 Animal and Plant Adaptation

2.1 Variation

Key Ideas

☐ 1 Variation is the range of differences found within a species.
☐ 2 Sexual reproduction is a means of maintaining genetic variation in a population and is also important in long-term evolutionary change.
☐ 3 Meiosis is a type of cell division that produces the gametes required for sexual reproduction from gamete mother cells within the sex organs.
☐ 4 In meiosis, independent assortment of chromosomes gives variation in the gametes.
☐ 5 In meiosis, crossing over allows the exchange of alleles between parental chromosomes to allow new combinations of parental variation to arise, increasing variation in the gametes.
☐ 6 Independent assortment and crossing over lead to the production of new phenotypes allowing variation to be maintained amongst offspring.
☐ 7 Gametes produced by meiosis are haploid, allowing the diploid number to be restored after fertilisation.
☐ 8 Meiosis occurs in two stages, the first and second meiotic divisions.
☐ 9 During the first meiotic division homologous chromosomes in a gamete mother cell pair. At this stage chiasmata may form between the pairs before the members of each pair are pulled apart by spindle fibres.
☐ 10 During the second meiotic division, the chromatids of each chromosome are split at their centromeres and are pulled apart to form four haploid gametes.
☐ 11 A dihybrid cross allows the inheritance of two pairs of contrasting characteristics to be studied. The inheritance of pink or white flowers along with the inheritance of tallness or dwarfness in pea plants is an example of dihybrid inheritance.
☐ 12 Phenotype ratios in offspring from dihybrid crosses can be predicted.
☐ 13 Alleles on the same chromosome are linked. Linked alleles are normally inherited together.
☐ 14 Dihybrid crosses involving linked alleles produce different phenotype ratios in offspring than those in which the alleles are not linked.
☐ 15 Linked alleles can be recombined after crossing over, altering the expected phenotype ratios in the offspring.

Key Ideas *continued* ➢

Key Ideas continued

- ☐ 16 The frequency of recombination between two linked alleles depends on the distance that separates them on the chromosome. The closer the alleles are, the less frequently they will be recombined.
- ☐ 17 In some organisms including humans, special chromosomes are involved in the inheritance of sex. These are the sex chromosomes.
- ☐ 18 In humans sex-linked alleles occur on the part of the X chromosome that is not homologous to the Y chromosome.
- ☐ 19 Mutant alleles have changes in their genetic information.
- ☐ 20 Mutant alleles occur randomly and infrequently but the rate of mutation is increased by mutagenic agents.
- ☐ 21 Mutation rate can be increased by chemical agents or by irradiation.
- ☐ 22 Mutation of a single allele involves the alteration of the DNA base type or sequence.
- ☐ 23 Substitution involves the removal of a base from an allele and its replacement with a different base. Inversion involves the changing of the positions of adjacent bases.
- ☐ 24 Deletion involves the removal of a base and insertion involves the addition of an extra base to the sequence.
- ☐ 25 Some mutations involve alterations to the number or structure of whole chromosomes.
- ☐ 26 Non-disjunction involves the failure of the spindle to part chromosomes during cell division and the subsequent change in the number of chromosomes in the daughter cells.
- ☐ 27 In polyploidy an individual possesses one or more entire sets of chromosomes in excess of the normal diploid number. Polyploidy results from complete non-disjuction of chromosomes.
- ☐ 28 In the breeding of crop plants, polyploidy has been used to increase the vigour of the plants.

Topic Notes

Meiosis and the dihybrid cross

Genetic variation is a term used for the range of differences found within a species. The differences often show up in the external appearance of the individuals. For example, the human population is characterised by a wide variation in colour of eye and hair. Genetic variation is maintained by a number of different processes.

Sexual reproduction itself is a means of enabling genetic variation to be maintained. This process is also important in long-term evolutionary change. Evolutionary change is usually gradual and leads to the development of new species over very long time periods.

The body cells of most adult organisms are diploid which means that they contain two sets of chromosomes (2n), one set inherited from each parent in sexual reproduction. Gametes on the other hand are haploid meaning that they contain only one set of chromosomes (n). The diploid condition is restored when two gametes fuse at fertilisation to produce a zygote. This is illustrated on the life cycle diagram.

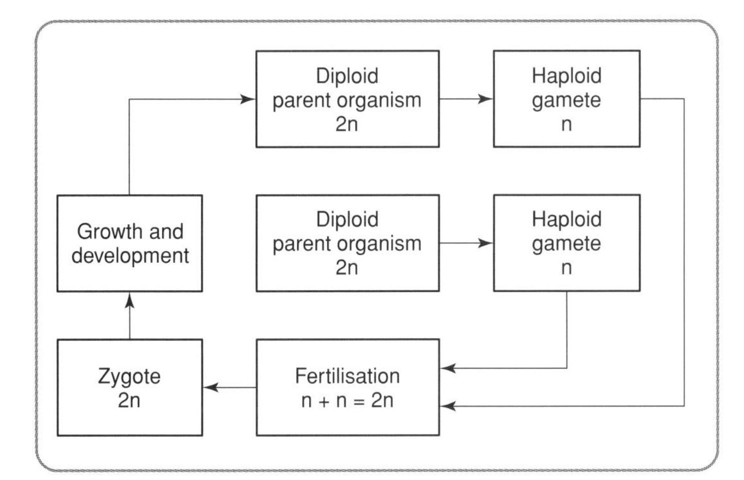

Figure 2.1 Life cycle diagram

Top Tip

In humans 2n = 46, in pea plants 2n = 14 and in *Drosophila* 2n = 8.

The maintenance of variation through sexual reproduction is connected to processes that occur during the production of gametes. Gametes are produced by a cell division called meiosis that occurs in the ovaries and testes of animals and in the ovaries and anthers of flowering plants. Meiosis involves the production of haploid (n) gametes from diploid (2n) gamete mother cells.

Meiosis occurs in two phases called the 1st and 2nd meiotic divisions. These phases result in the production of haploid gametes from gamete mother cells. An outline of meiosis is given in the following set of diagrams.

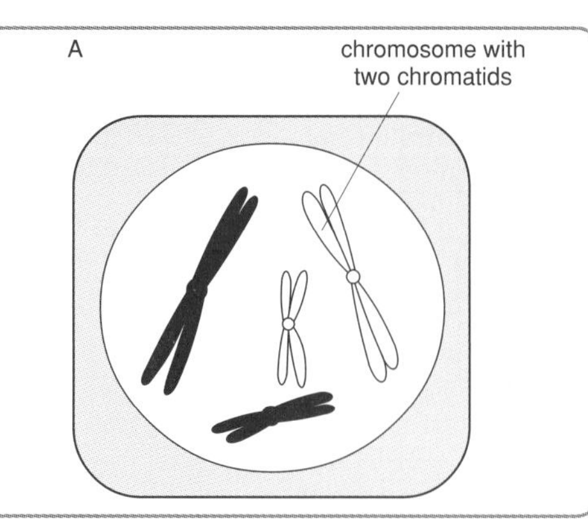

A Gamete mother cell with 2n = 4 chromosomes, each chromosome seen to consist of two chromatids

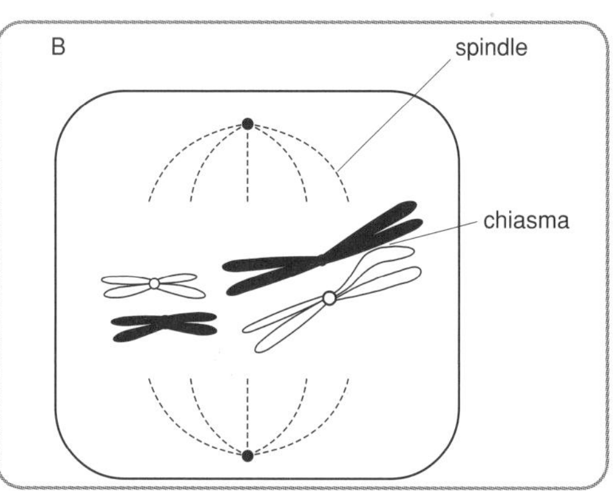

B Independent assortment occurs as homologous chromosomes pair together, crossing over can occur at chiasmata, a spindle forms and the pairs move to its equator

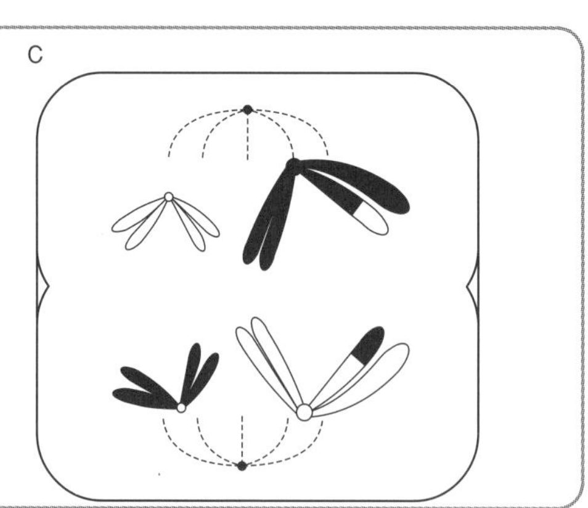

C Homologous chromosome pairs are pulled apart to opposite ends on the spindle

Figure 2.2 First meiotic division

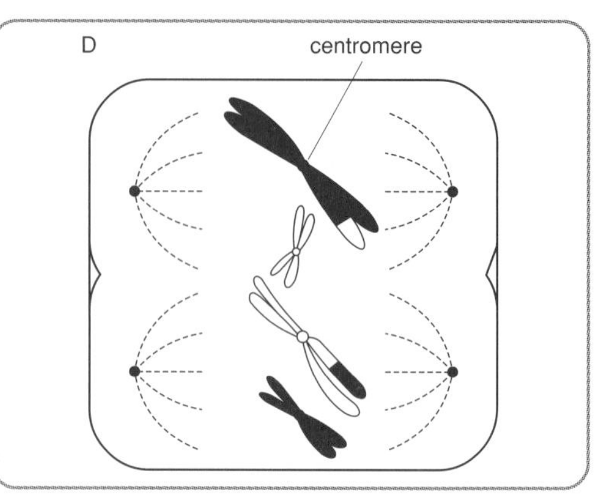

D Two new spindles form and the chromosomes are pulled to the equators

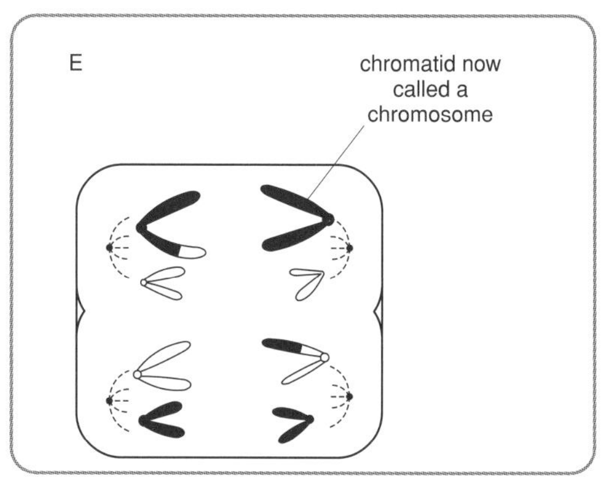

E The centromere of each chromosome splits and the chromatids, now called chromosomes, move to the opposite ends of each spindle

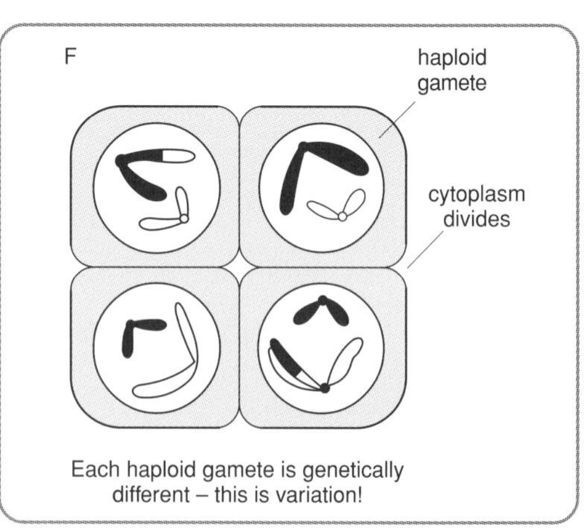

F The cytoplasm divides resulting in the production of four haploid gametes

Figure 2.3 Second meiotic division

Top Tip

1st meiotic division separates homologous pairs to give two haploid cells.

2nd meiotic division separates pairs of chromatids giving four haploid gametes.

As can be seen in diagram F, the combinations of chromosomes in the gametes are varied. This variation has been caused by **two** processes that occurred during the 1st meiotic division as shown in diagram B.

1. **Crossing over** between homologous chromosomes, as shown in diagram B, can occur leading to the recombination of groups of alleles and resulting in genetically different gametes.
2. **Independent assortment** of chromosomes arises as the homologous pairs line up randomly. Digram B shows how the pairs lined up in this example, but the diagram below shows an alternative that would give other genetic possibilities and more variation.

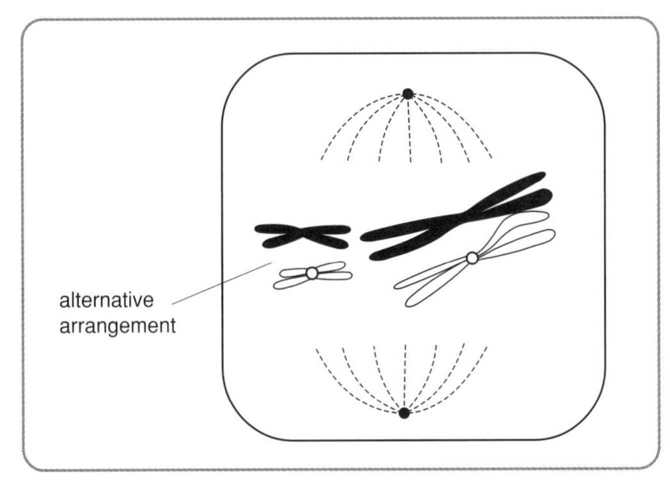

Figure 2.4 An alternative arrangement at stage B

Top Tip

Engage in this activity!

Have a go at drawing the chromosome combinations in the four haploid gametes that would be produced from the alternative arrangement of pairs shown in Figure 2.4.

Both **independent assortment** and **crossing over** are means of producing new phenotypes. This can be illustrated by the study of genetic crosses involving the inheritance of two pairs of contrasting characteristics. Such crosses are called dihybrid crosses and some classic examples were described in the mid-nineteenth century by Gregor Mendel an Austrian monk. Mendel worked on garden pea plants and chose to study various genes occurring on different chromosomes of the plant. The best known example involved the inheritance of the plants' height which could be tall – the plants growing to about 3 metres in height, contrasting with dwarf – the plants only growing to about 0.5 metres resulting in a much bushier appearance. Additionally petal colour could be pink or white.

To fully understand the steps and significance of a dihybrid cross, some vocabulary is essential and you will need these words for your exam!

Key Words *and* ***Definitions***

It's better to learn these words as a group since many of the definitions rely on knowing other definitions.

Gene A single piece of genetic information coding for one characteristic – in pea plants there is a **gene** coding for petal colour.

Allele A form of a gene – in pea plants the gene for petal colour has two different **alleles** giving either pink or white petals. Alleles are usually written as capital (C) or small (c) letters.

Homozygous Organisms **homozygous** for a characteristic have two identical alleles for the characteristic – a pea plant homozygous for petal colour has two identical petal colour alleles, both pink CC or both white cc.

Heterozygous Organisms **heterozygous** for a characteristic have two different alleles for the characteristic – a pea plant heterozygous for petal colour has two different alleles for petal colour, one pink and one white Cc.

Dominant **Dominant** alleles always show up in the phenotype of an organism and are given capital letters as symbols – a pea plant that inherits a copy of the allele C for pink petals will have pink petals even if the other allele it inherits for petal colour is c for white petals since pink is dominant to white.

Recessive **Recessive** alleles only show up in the phenotype of an organism if the organism inherits two copies of the allele – a pea plant can only have white petals if it inherits two copies of the allele c.

Phenotype The **phenotype** of an organism is the physical expression of its genes – a pea plant might have pink flowers or white flowers.

Genotype The **genotype** of an organism is a statement of its alleles for a particular characteristic usually given as symbols – a pea plant could have the genotype CC if it were homozygous for pink petal colour.

In pea plants, tall (T) is dominant to dwarf (t) and pink petals (C) is dominant to white petals (c). In a dihybrid cross, homozygous tall plants with pink flowers are crossed with homozygous dwarf plants with white flowers. The heterozygous offspring form the F1 generation and are all tall with pink flowers. Members of the F1 are crossed together. Their offspring form the F2. This can be set out as shown in the next diagram.

Parents Gametes	homozygous tall pink TTCC TC	X	homozygous dwarf white ttcc tc
F1	all heterozygous tall pink TtCc		
F1 cross Gametes	heterozygous tall pink TtCc TC, Tc, tC, tc	X X	heterozygous tall pink TtCc TC, Tc, tC, tc

The F2 Punnet Square diagram is a grid that can be used to show all possible fertilisations from a dihybrid cross.

Gametes	TC	Tc	tC	tc
TC	TTCC	TTCc	TtCC	TtCc
Tc	TTCc	TTcc	TtCc	Ttcc
tC	TtCC	TtCc	ttCC	TtCc
tc	TtCc	Ttcc	ttCc	ttcc

If the phenotypes of the offspring given in the table are counted, the simple whole number ratio gives the expected offspring phenotype ratio for the cross.

Top Tip

The F_2 phenotype shows the following ratio:

9 tall pink : 3 tall white : 3 dwarf pink : 1 dwarf white

Use a coloured crayon to shade all the boxes containing genotypes which would give tall pink plants. Use 3 different colours to do the same for the three other phenotypes – they come out in an obvious pattern.

It has been shown by experiment that dihybrid crosses with various organisms very often follow this pattern giving a 9 : 3 : 3 : 1 phenotype ratio in the F2.

The expected ratios do not always occur. One reason is that the numbers of offspring produced by a cross may be too small for reliable ratios to be produced. Also, the alleles chosen in a cross may be linked on the same chromosome.

Linkage and crossing over

Linked genes occur on the same chromosome. Dihybrid crosses with linked alleles give a different F2 phenotype ratio.

In the fruit fly *Drosophila*, the allele for normal wings (W) is dominant to the allele for vestigial wings (w) and the allele for grey body (G) is dominant to the allele for black body (g).

If a fly homozygous for normal wings and grey body is crossed with a fly homozygous for vestigial wings and black body all the offspring are heterozygous for normal wings and grey body. However, if these F1 flies are crossed, the offspring do not produce the expected 9 : 3 : 3 : 1 ratio. This is because the alleles are linked on the same chromosome.

This cross is shown on the following diagram with drawings of the chromosomes carrying the linked genes included.

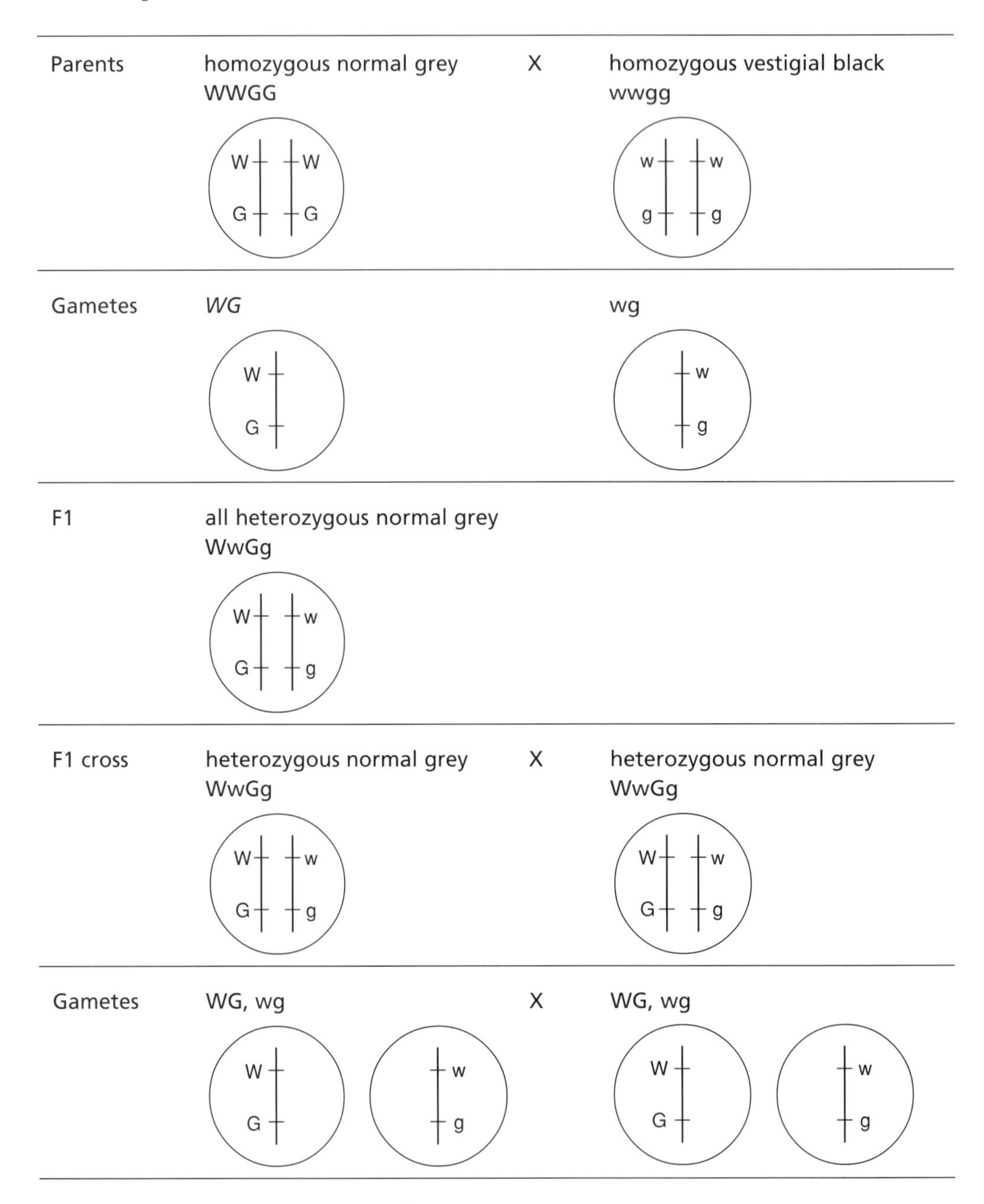

F2 Punnet Square diagram

gametes	WG	wg
WG	WWGG	WwGg
wg	WwGg	wwgg

As can be seen from the punnet square, the offspring phenotype ratio has been affected by the linkage pattern and works out as shown.

3 normal grey : 1 vestigial black

However, sometimes you may be told in a question that, although a cross involved linkage, the expected 3:1 ratio did not occur. This could be due to lack of a sufficient number of offspring for reliability or because of crossing over. The following figure shows the gametes possible with and without crossing over between the *Drosophila* alleles from the cross above.

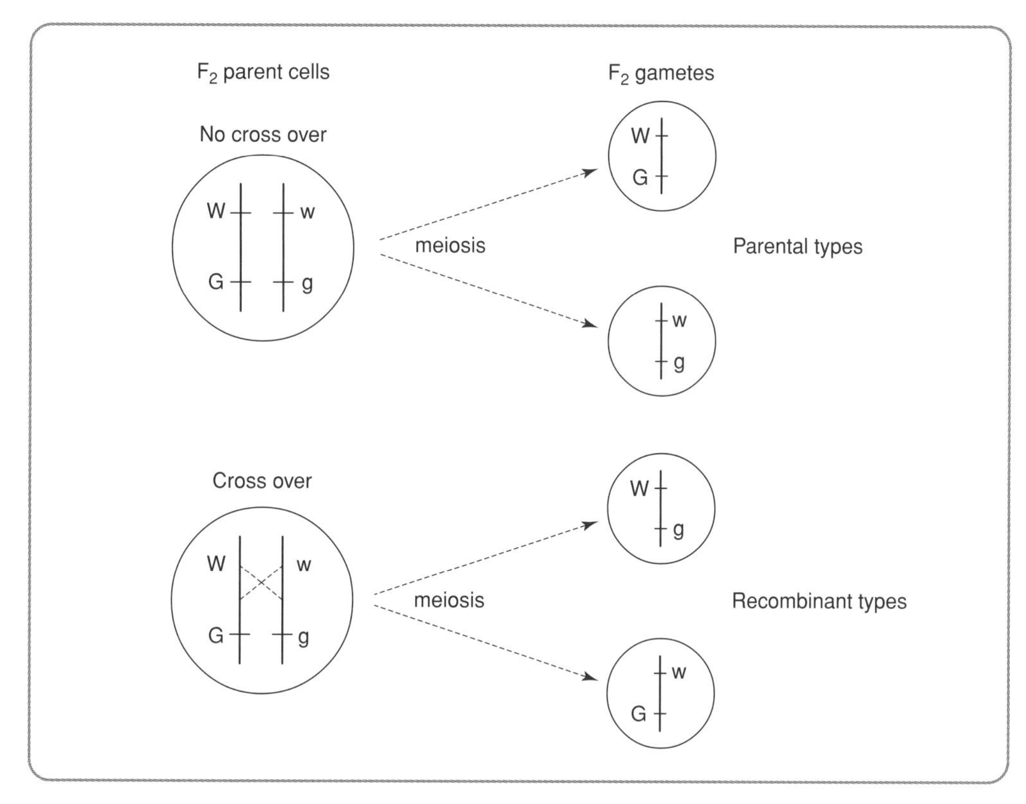

Figure 2.5 Possible gametes with and without crossing over

Crossing over during the 1st meiotic division can cause the break up of linkage groups and the formation of recombinant gametes. If these gametes are involved in fertilisation, **recombinant** offspring can be formed. In the *Drosophila* cross above, the F2 generation will contain a small number of recombinant offspring with either normal wings and black bodies or vestigial wings and grey bodies. The vast majority of the offspring will be **parental** types with either long wings and grey bodies or vestigial wings and black bodies. Figure 2.5 shows the gametes produced without and with crossing over between the wing size and body colour alleles.

Recombination Frequencies

The frequency of occurrence of recombination is related to the position of the alleles on the chromosome. The further two alleles are separated, the greater the frequency of recombination between them. The following figure shows this idea applied to three alleles A, B and C on a chromosome.

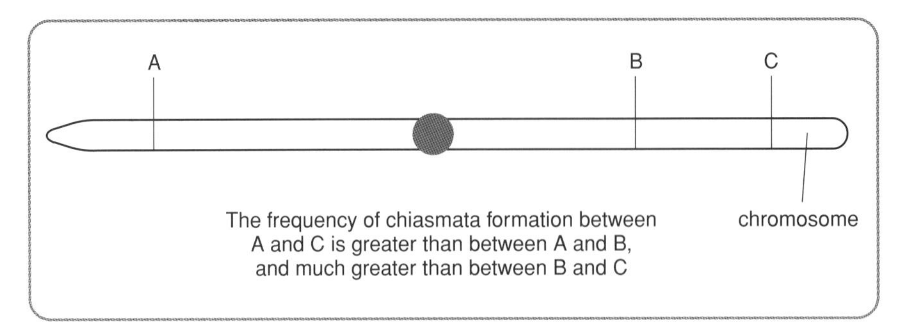

Figure 2.6 Frequency of occurrence of recombination

The recombination of linked genes during meiosis is another source of variation provided by the process.

Sex determination

In humans, sex is determined by the sex chromosomes. Females have two X chromosomes and males have one X chromosome and one Y chromosome. Figure 2.7 shows that the X chromosome has a section that is not homologous with the Y chromosome. The genes carried on this section are sex-linked. Females have two sex-linked alleles of each gene whereas males only have one.

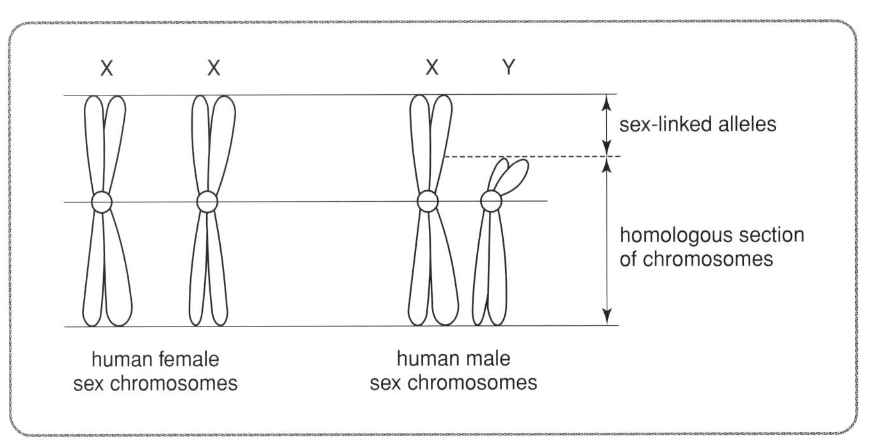

Figure 2.7 Human sex chromosomes

In humans, red–green colour deficiency is sex-linked. The normal allele (R) is dominant to the allele (r) that causes the deficiency. Figure 2.8 shows the inheritance of red–green colour deficiency. In this family, the mother is a carrier of the deficiency. She has the deficiency allele r on one of her X chromosomes, but the dominant allele R on her other X chromosome gives her a normal phenotype.

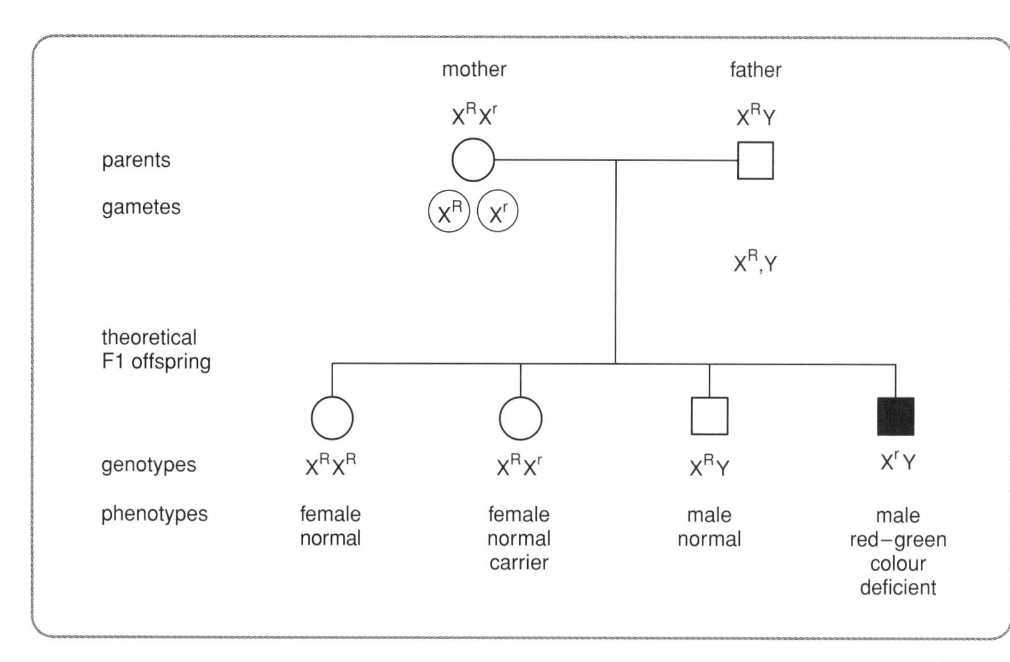

Figure 2.8 Inheritance of red–green colour deficiency

Mutation

Mutation is a source of entirely new variation in a population. Mutations are random events that cause changes to the genetic information of an organism. They can affect either single alleles or entire chromosomes.

Mutations occur at low frequency but the rate of mutation can be increased by mutagenic agents. These agents can either be chemicals, such as the drug colchicine and the chemical weapon mustard gas, or irradiation such as X-rays and ultraviolet (UV) light.

Top Tip

Remember **ROLF**!

Random **O**ccurrence, **L**ow **F**requency

Mutations of single genes are changes in the base sequence of the gene – a bit like spelling mistakes in a long word. Bases can be substituted, inverted, deleted or inserted.

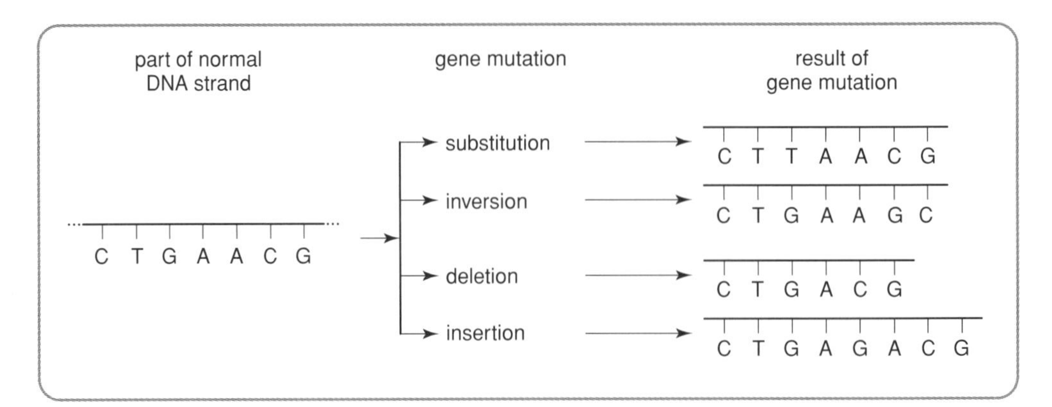

Figure 2.9 Gene mutation

Top Tip

Remember **DIGSI!**

Deletion, **I**nsertion (**G**ene mutations) **S**ubstitution, **I**nversion

Many candidates confuse inversion and insertion and so lose marks.

Substitution and inversion have less drastic effects on an allele as they make only one or two differences to the triplets in the base sequence and, therefore, to the amino acid sequence of the protein coded for. Deletion or insertion can have drastic effects on the allele since they often affect many of the bases in the sequence and therefore many amino acids in the protein produced, making the triplets of protein non-functional.

Changes in the number of chromosomes arise by non-disjunction in which one or more chromosome pairs fail to move apart because of spindle failure during cell division. This results in cells with more or fewer chromosomes than normal.

In humans, failure of chromosome pair 21 to separate during meiosis can result in some gametes with an extra copy of this chromosome. If such a gamete is fertilised by a normal gamete the resulting zygote has three copies of chromosome 21. The offspring will have a condition called Down's Syndrome.

Complete non-disjunction can produce cells with entire extra sets of chromosomes. Polyploidy is a term used to describe the condition in which an organism has one or more sets of chromosomes in excess of the normal diploid number.

Polyploidy has been used to advantage in crop production. Many modern crops such as bread wheat are polyploid. In these cases, polyploidy has conferred increased vigour such as fast growth, resistance to disease and high yields (to the crop species).

Some mutations involve changes in the structure of one chromosome in a set. These include duplication, inversion, translocation and deletion as shown in Figure 2.10.

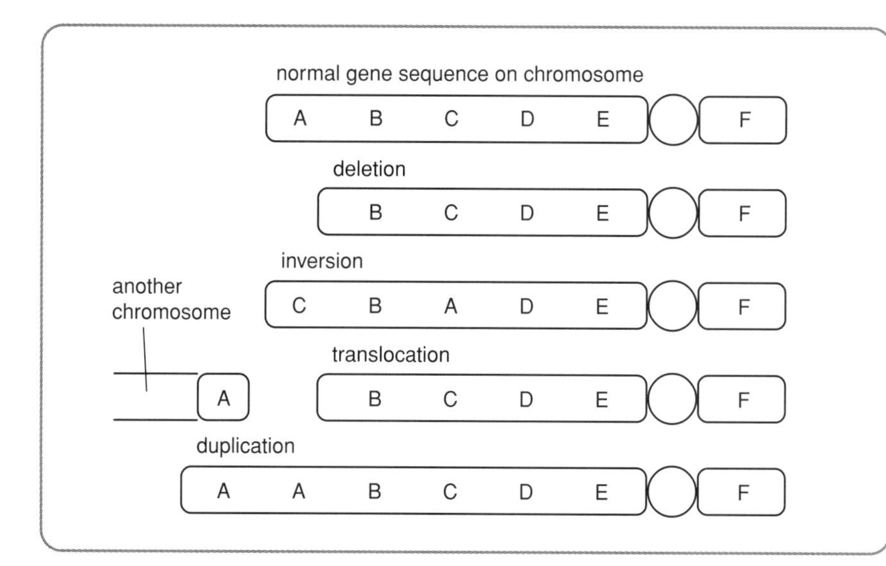

Figure 2.10 Changes in structure of a chromosome

Top Tips

Remember **DICTD!**

Duplication, **I**nversion (**C**hromosome mutations) **T**ranslocation, **D**eletion

Many candidates, understandably, confuse deletion and inversion of **chromosomes** with deletion and inversion of **single genes**. Make sure you read any question involving these terms with care!

Exam Questions

Marks

SBQ 1 The diagrams show five stages in meiosis.

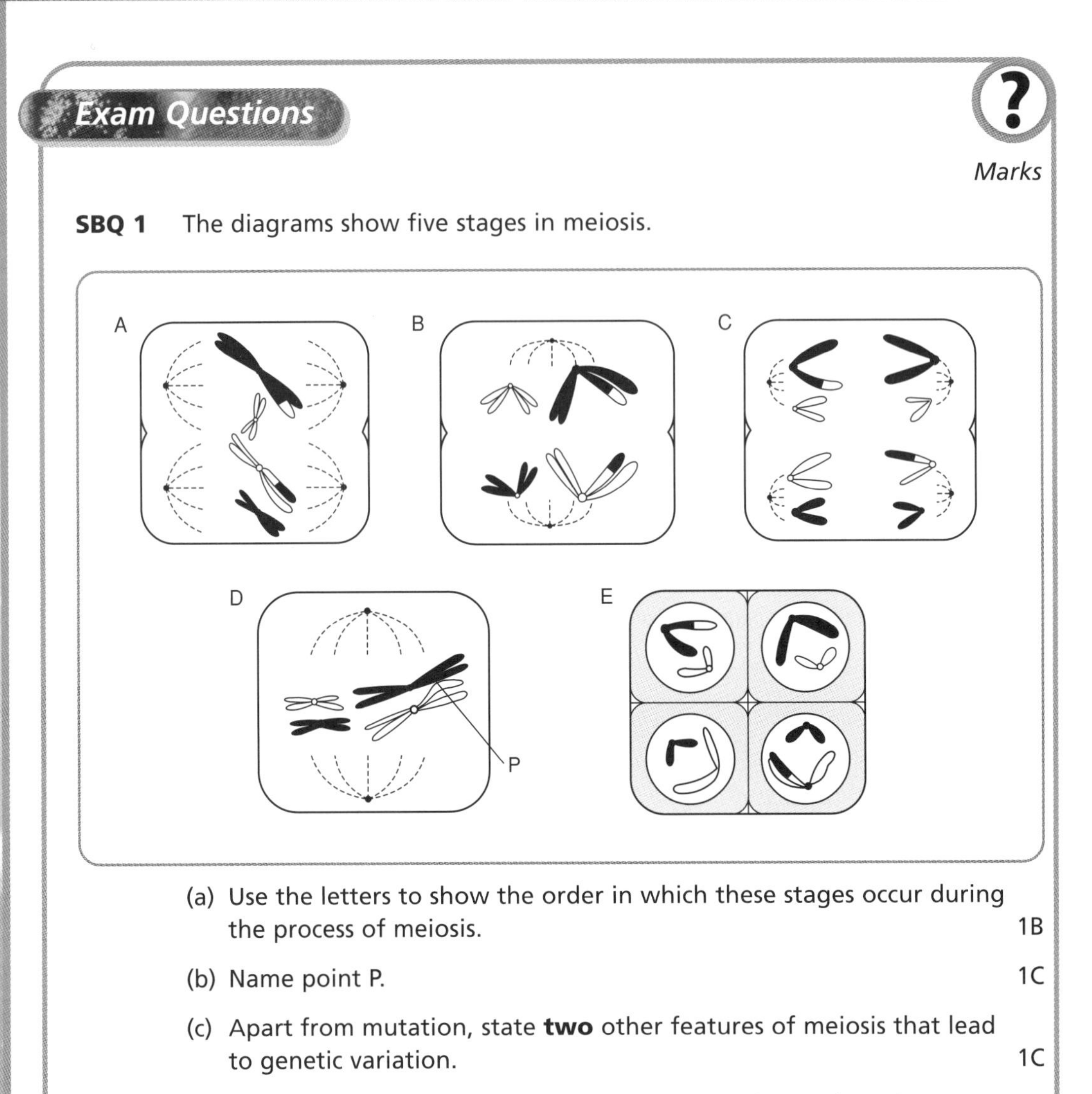

(a) Use the letters to show the order in which these stages occur during the process of meiosis. 1B

(b) Name point P. 1C

(c) Apart from mutation, state **two** other features of meiosis that lead to genetic variation. 1C

SBQ 2 In a certain organism, the genes J, K, L and M are located on the same chromosome. The table shows the frequency of recombination between the different genes.

Gene pairs	Frequency of recombination %
J and K	8
J and L	14
K and L	22
K and M	4
M and J	12

Give the order of the genes relative to each other on this chromosome. 1B

***Exam Questions** continued* ➢

Exam Questions continued

SBQ 3 In cocker spaniel dogs, the allele for black coat (B) is dominant to red coat (b) and the allele for plain coat (C) is dominant to spotted coat (c).

A male with plain black coat was mated to a female with plain red coat. Of the six offspring produced, two had plain black coats, two had plain red coats, one had a spotted black coat and one had a spotted red coat.

(a) State the genotypes of the parents. 1A

(b) Give the **theoretically** expected ratio of phenotypes in the offspring from this cross. 1A

SBQ 4 In humans, the allele for red–green colour deficiency (d) is sex-linked and recessive to the normal allele (D).

The diagram below shows how the condition was inherited in a family.

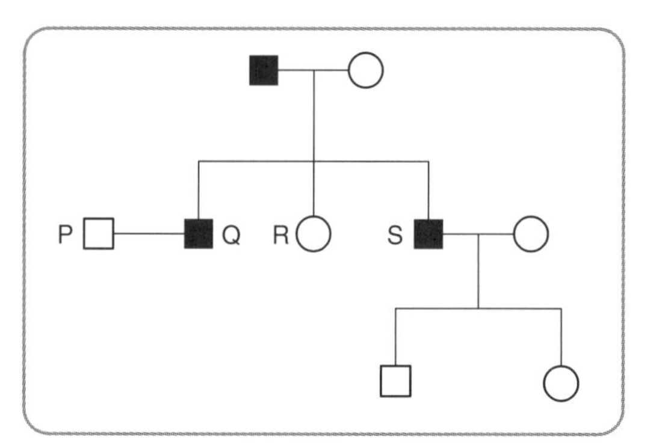

(a) State the genotypes of individuals R and S. 1C

(b) If individuals P and Q have a daughter, state the percentage chance that she will show red–green colour deficiency. 1B

(c) Explain why individual Q shows the condition even though her mother was not affected. 1A

SCQ 1 Give an account of gene mutation under the following headings:

(i) the characteristics of mutant alleles and the effects of mutagenic agents 3

(ii) the effects of named gene mutations on amino acid sequences in proteins. 7

(10)

In question SCQ 2, ONE mark is available for coherence and ONE mark is available for relevance.

Exam Questions continued ➢

Exam Questions *continued*

SCQ 2 Give an account of chromosome mutation with reference to changes in chromosome number and structure and to polyploidy in crop plants.

(10)

Answers on pages 136–8

2.2 Selection and Speciation

Key Ideas

- ☐ 1 A species is a group of organisms that can interbreed to produce fertile offspring.
- ☐ 2 A species has a common gene pool.
- ☐ 3 New species of living organisms have evolved as a result of isolation, different mutations followed by natural selection.
- ☐ 4 Natural selection involves survival of organisms best suited to their environment.
- ☐ 5 A species can be split into separate populations by isolating mechanisms that can be geographical, ecological or reproductive.
- ☐ 6 Isolating mechanisms act as barriers to gene exchange and prevent the isolated populations from normal interbreeding.
- ☐ 7 Mutations arising in one isolated population cannot be passed to other isolated groups.
- ☐ 8 Natural selection operates separately on each isolated population, causing differences to develop between them as successful members of the populations pass their favourable characteristics on to their offspring.
- ☐ 9 After long periods of time, differences in the gene pools of the isolated populations become so great that members of the separate populations cannot interbreed even if brought together by the breaking down of the isolating barrier.
- ☐ 10 Speciation is the formation of two or more species from the original one.
- ☐ 11 Speciation can involve adaptive radiation.
- ☐ 12 In adaptive radiation, a common ancestor evolves to produce several new species that are different from each other depending on their habitat and, in some cases, their method of feeding.
- ☐ 13 Some organisms have evolved at unusually high speeds because certain mutations have been very strongly favoured by natural selection. Examples include antibiotic resistant strains of bacteria and the melanic form of the peppered moth.
- ☐ 14 Some species in danger of extinction have been conserved through wildlife reserves, captive breeding programmes and cell banks.
- ☐ 15 Conservation of endangered species helps to maintain genetic diversity.
- ☐ 16 Various crop and domesticated animal species have been created by artificial selection by humans.
- ☐ 17 Selective breeding aims to enhance desirable characteristics by choosing individuals showing these characteristics as parents.

Key Ideas *continued* ➢

Key Ideas continued

- ☐ 18 Hybridisation involves cross-breeding two separate species to allow the combination of their desirable characteristics in their offspring.
- ☐ 19 Hybridisation is limited as species may be sexually incompatible and will not cross-breed or, if they do, their offspring are infertile.
- ☐ 20 Somatic fusion can be used to overcome sexual incompatibility between two different species of plant by removing the cell walls from somatic cells with cellulase and fusing the resulting protoplasts.
- ☐ 21 Infertility in hybrids can be overcome by artificially encouraging polyploidy.
- ☐ 22 Genetic engineering has contributed to the development of new varieties of organisms to human advantage. In genetic engineering, the genetic information of one species is combined into the genetic information of another.
- ☐ 23 The location of genes or groups of genes on chromosomes is essential for genetic engineering. This is achieved using gene probes or by the recognition of characteristic banding patterns on stained chromosomes.
- ☐ 24 After location, the genes can be cut from their chromosome using restriction endonuclease enzymes.
- ☐ 25 Ligase enzymes are used to seal the genes into the genomes of other organisms such as bacteria.
- ☐ 26 The bacteria can be cultured and the products of the activity of the inserted genes isolated and purified. Examples of genetically engineered products include human growth hormone and insulin.

Topic Notes

Natural selection

A species is a group of interbreeding organisms that produce fertile offspring. Some species can interbreed with other species but the offspring are infertile.

Evolution of new species from one ancestral species involves vast periods of time. It requires **isolation** of populations, variation produced by **mutations**, and the process of **natural selection**.

Isolating mechanisms are barriers to gene exchange between two sub-populations of a single species. The main types of isolating mechanisms you need to remember for your exam are geographical, reproductive and ecological.

Top Tip

Remember **GERM!**

Geographical, **E**cological, and **R**eproductive **M**echanisms

Geographical barriers are physical and cannot be crossed by members of sub-populations. These include mountain chains, oceans and rivers.

Ecological barriers are habitat differences. The sub-populations live in different habitats in which other sub-populations cannot survive.

Reproductive barriers are differences connected with breeding. One sub-population of flowering plant may produce flowers at a different season from another sub-population.

If two sub-populations of a single species become separated by an isolating mechanism, different **mutations** occurring in each will not be passed between them.

These **mutations** are new variations that arise randomly in sub-populations populations. The majority of these are either harmful or of no significance to individual organisms. Occasionally a mutation that is beneficial to the individual organism occurs.

Top Tip

It's important to emphasise that the population is split into **sub**-populations and that mutations are **different** in each sub-population. Many candidates fail to emphasise these points and lose marks.

A beneficial mutation occurring in one sub-population may allow the mutant to survive to breed and pass on the beneficial characteristic to offspring making the mutation gradually become more common in that sub-population. This is **natural selection** – the survival to breed of those organisms most suited to their environment. Since an isolating barrier separates the sub-populations, gene exchange between them cannot occur and genetic differences caused by mutations will accumulate. Over long periods of time, each sub-population will gradually evolve into a new species. If these new species were allowed to interbreed by the removal of the isolating barrier, they could not do so successfully as their genetic material would be so different.

Adaptive radiation is the evolution of a group of new species from a common ancestor. The new species are often characterised by marked differences in an important characteristic. Figure 2.11 shows the heads of four species of Darwin's finches that have evolved by adaptive radiation on the Galapagos Islands isolated in the Pacific Ocean. The important

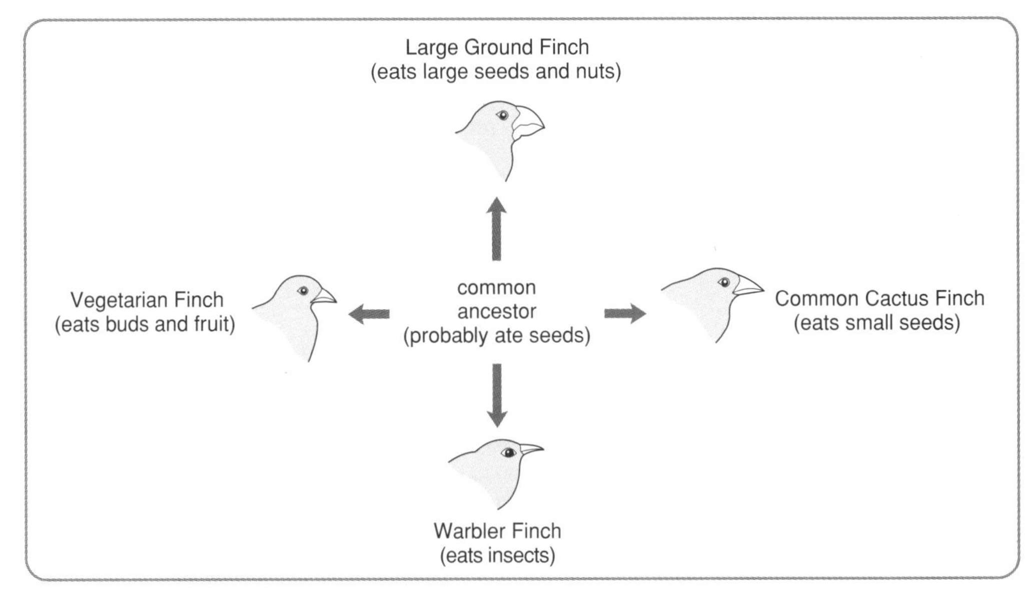

Figure 2.11 Four species of Darwin's Finch and their main food

characteristic that is clearly different is the size and shape of the birds' beaks suggesting that they are adapted to different food sources.

Sometimes evolution can proceed at high speed if there is an environmental change which drastically affects survival. One example of this is the emergence of two varieties of the peppered moth (*Biston betularia*) in Britain during and after the industrial revolution in the mid-nineteenth century. The main events of this are shown in Figure 2.12.

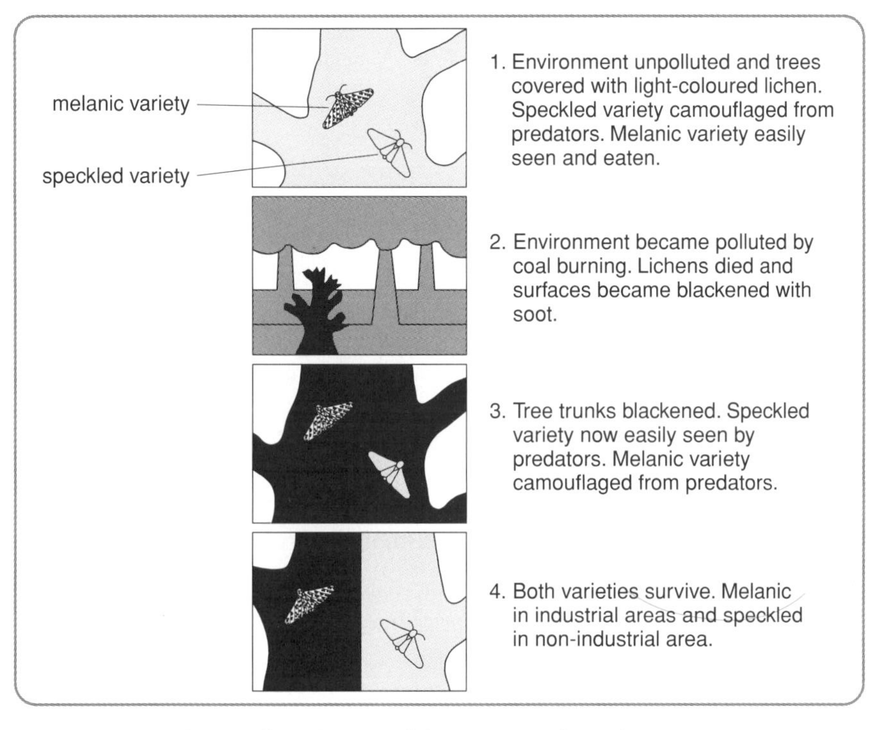

Figure 2.12 High speed evolution of the peppered moth

Another example of high speed evolution is the emergence of strains of antibiotic resistant bacteria during and after the extensive use of antibiotic drugs in the mid-twentieth century. Figure 2.13 shows the main events in this example.

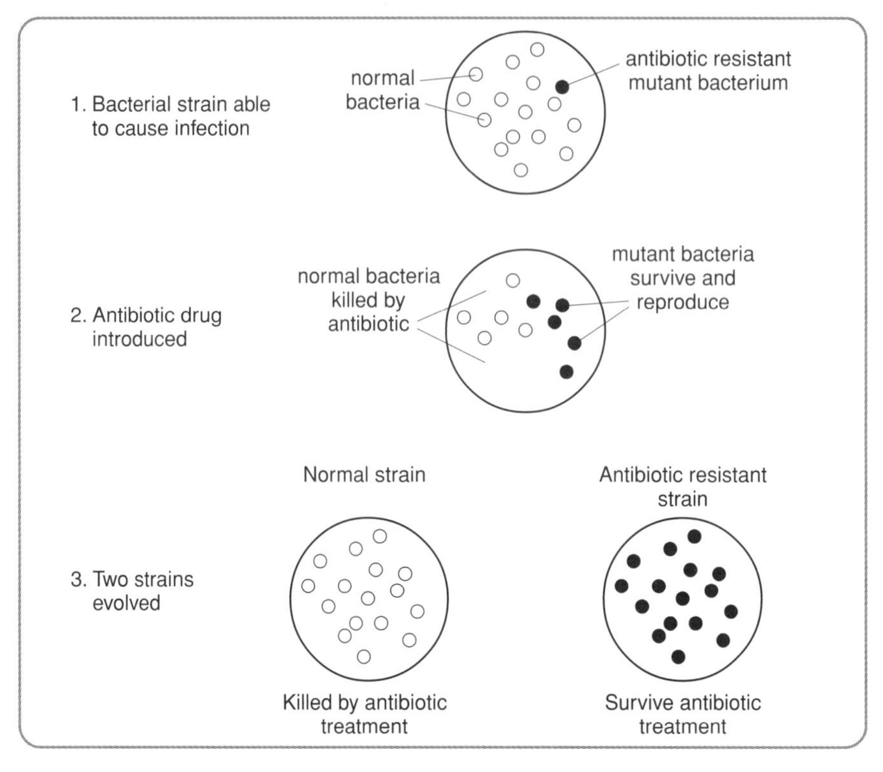

Figure 2.13 High speed evolution of antibiotic resistant bacteria

Some species are in danger of extinction. In order to maintain genetic diversity, it has become necessary to conserve some of these species through the use of wildlife reserves, captive breeding programmes and cell banks.

Artificial Selection

Artificial selection is similar to natural selection but is practised by humans to improve domesticated animals and crop plants. By choosing parent organisms with desired characteristics it is possible to improve an organism over time. Cattle varieties with high milk yields have been produced by continual selection of naturally high milk yielding cows crossed with bulls from a pedigree of high milk yield.

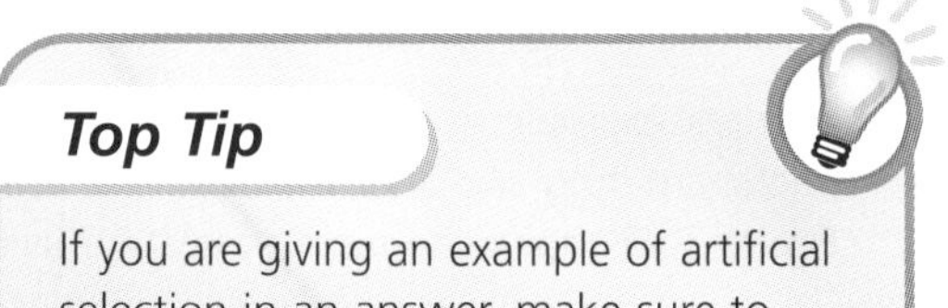

Top Tip

If you are giving an example of artificial selection in an answer, make sure to emphasise the improved characteristic, for example, milk **yield** in cattle.

Artificial selection can only draw on the natural variation within a species. Breeders sometimes want to combine desirable characteristics of two different species together. This can be done by hybridisation, the crossing of two different species. Hybridisation is more easily done in plants and, remember, any offspring produced are infertile. In plants, infertile offspring can reproduce by asexual vegetative methods, or drugs can be used to encourage polyploidy which can sometimes make the hybrids fertile.

Some plant species are sexually incompatible which means that they will not hybridise and no offspring at all can be produced. Sexual incompatibility between plant species can be overcome using somatic fusion. The stages involved in somatic cell fusion to produce a new variety of potato plant are shown in Figure 2.14. It shows the main stages in somatic fusion being used to create a high-yielding potato plant with disease resistance.

Top Tip

A **somatic** cell is a body cell not a gamete and **fusion** is a joining process a bit like fertilisation.

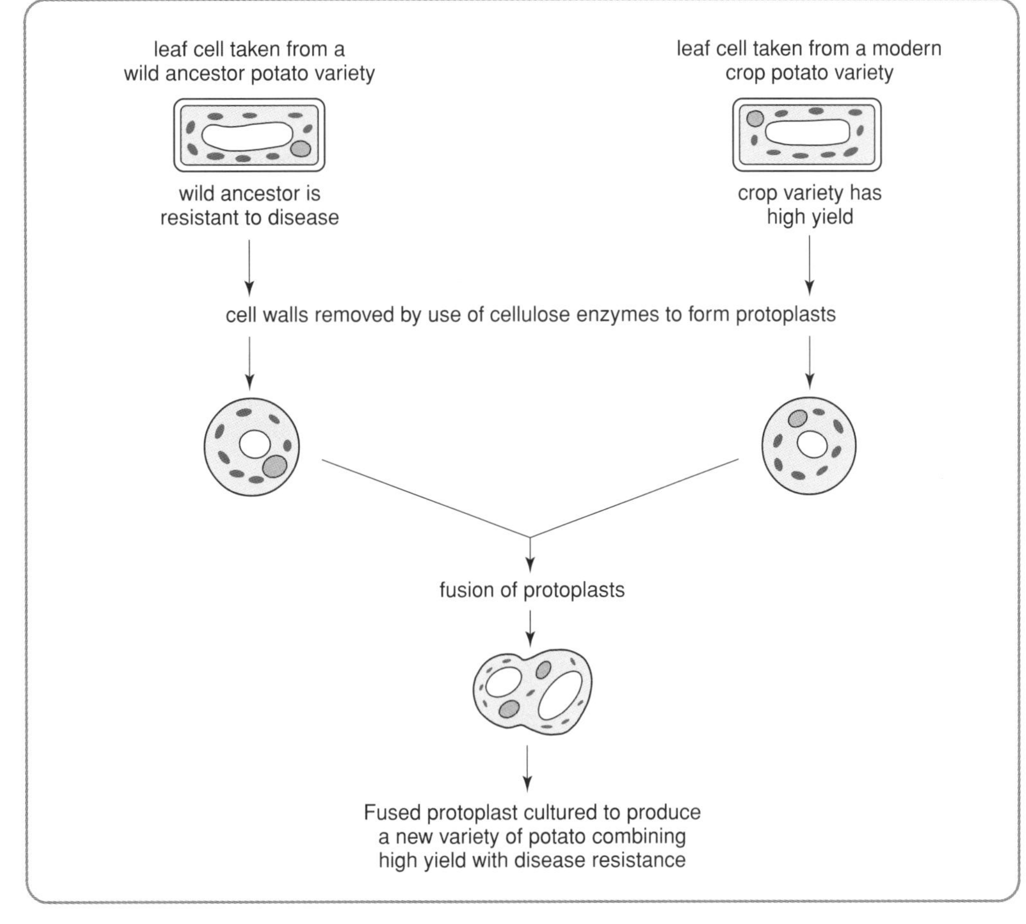

Figure 2.14 Somatic fusion to create a disease resistant pototo plant

If the genetic information from two extremely different species is to be combined, techniques of genetic engineering are used. It has been possible to combine human genes into the genomes of bacteria then culture the engineered bacteria and extract and purify the product of the human gene. This method has been used in the manufacture of human insulin and human growth hormone. Figure 2.15 shows the main stages in the process of genetic engineering.

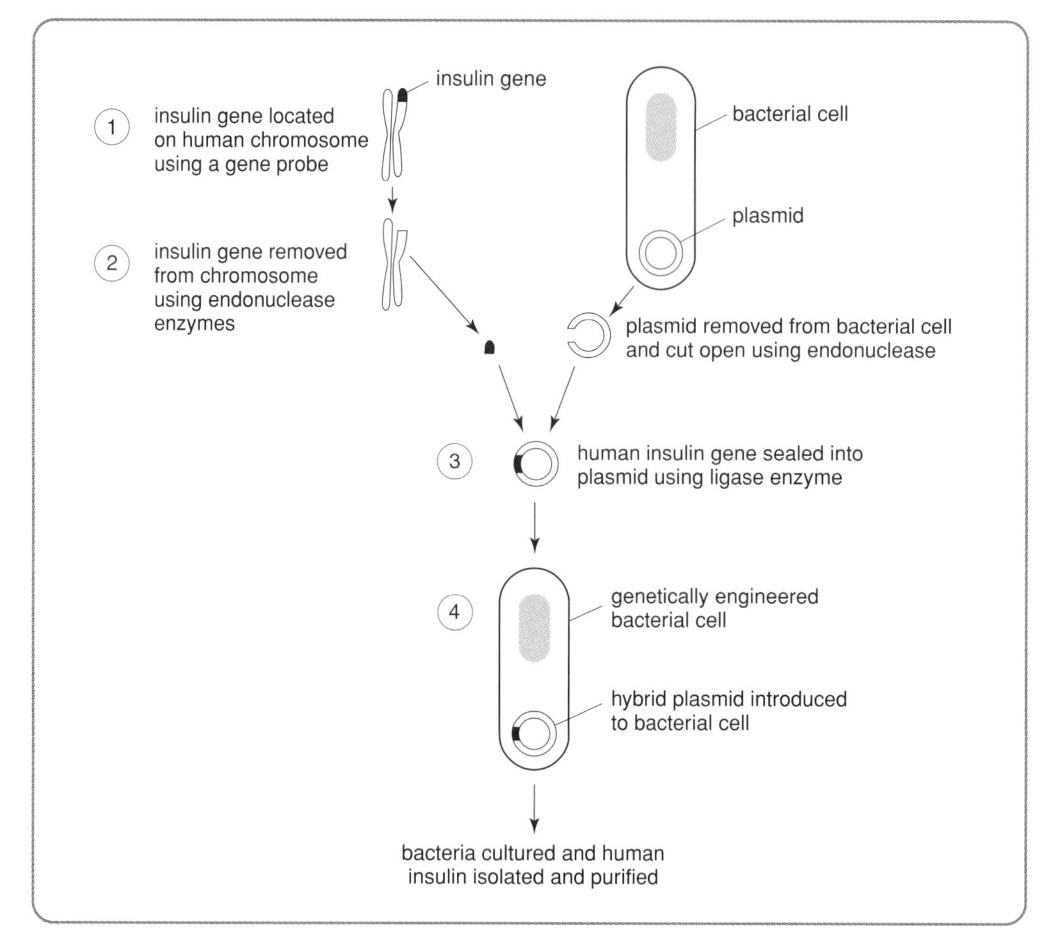

Figure 2.15 Producing insulin by genetic engineering

The advantages of this type of genetic engineering are that it provides unlimited quantities of the protein, high levels of purity and reduced risk of rejection.

Exam Questions

Marks

SBQ 1 The Galapagos Islands are isolated in the Pacific Ocean.
Different species of finch have evolved on these islands.
The heads of four of the finch species are shown.

***Exam Questions** continued* ➢

Exam Questions continued

(a) Explain how the information given about the Galapagos finches supports the statement that they have evolved by adaptive radiation. 1A

(b) What further evidence would be needed to be certain that the finches provide an example of adaptive radiation. 1B

(c) The finches have evolved in geographical isolation.
Name **two** other types of isolating barrier involved in the evolution of new species. 1C

(d) What evidence would confirm that each of the Galapagos finches was a separate species. 1C

(e) Some of the finch species face extinction.
Give **two** methods by which species can be conserved. 2C

SBQ 2 The diagram shows some stages in the genetic engineering of a bacterial cell to seal a human gene into the bacterial genome.

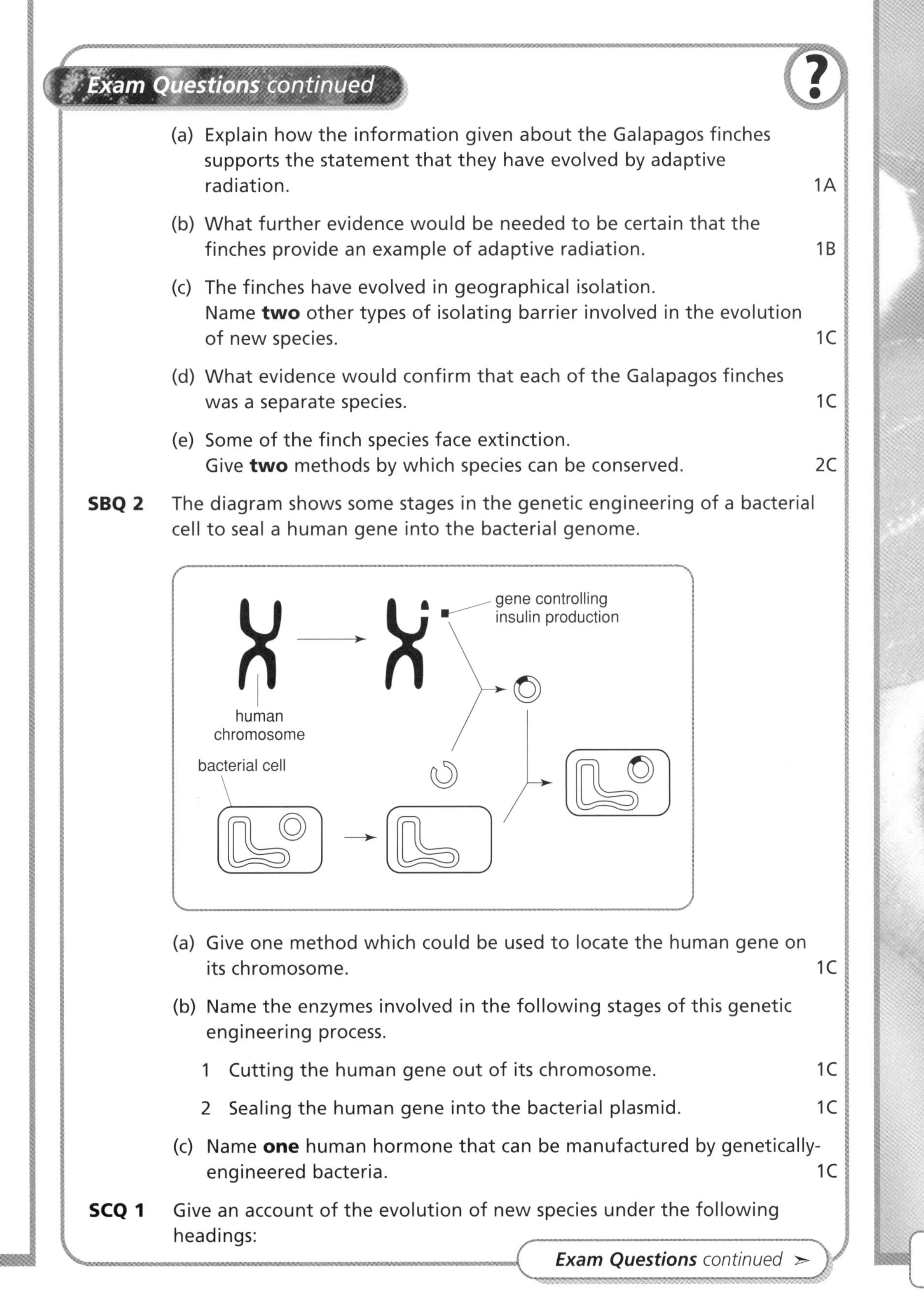

(a) Give one method which could be used to locate the human gene on its chromosome. 1C

(b) Name the enzymes involved in the following stages of this genetic engineering process.

1 Cutting the human gene out of its chromosome. 1C

2 Sealing the human gene into the bacterial plasmid. 1C

(c) Name **one** human hormone that can be manufactured by genetically-engineered bacteria. 1C

SCQ 1 Give an account of the evolution of new species under the following headings:

Exam Questions continued ➢

Exam Questions continued

(i) isolating mechanisms 4

(ii) the effects of mutations and natural selection 6

(10)

In question SCQ 2, ONE mark is available for coherence and ONE mark is available for relevance.

SCQ 2 Give an account of selective breeding, hybridisation and somatic fusion in the production of new varieties of plants. (10)

Answers on pages 138–40

2.3 Animal and Plant Adaptations

Key Ideas

☐ 1 Adaptations are features of an organism that help it to survive.

☐ 2 Freshwater fish have adaptations to help them osmoregulate. They have kidneys with many, large glomeruli that give a high filtration rate allowing the production of large volumes of dilute urine. They also have chloride secretory cells in their gills that actively transport salts into their blood.

☐ 3 Saltwater bony fish have kidneys with few, small glomeruli that give a low filtration rate and the production of small volumes of concentrated urine. They also have chloride secretory cells that actively excrete salts from their blood to the surrounding sea.

☐ 4 Migratory fish such as eels and salmon are able to alter their kidney functions and the direction of action of their chloride secretory cells depending on whether they are in fresh or salt water.

☐ 5 Desert mammals have both physiological and behavioural adaptations for conservation of water.

☐ 6 Physiological adaptations include dry mouths and nasal passages, long loops of Henle in the kidney that allow increased reabsorption of water from their glomerular filtrate, and the ability to produce dry faeces.

☐ 7 Behavioural adaptations include nocturnal habits and the use of moist burrows during the heat of the day.

☐ 8 The transpiration stream refers to the movement of water from the soil, through root hairs and cortex cells through the xylem to leaves and out to the atmosphere. Transpiration is the evaporation of water from stomata.

☐ 9 The column of water in the xylem is maintained using the forces of cohesion and adhesion of water molecules.

☐ 10 Various factors such as temperature, humidity, light and air movements affect transpiration rate.

***Key Ideas** continued* ➢

Key Ideas continued

- ☐ 11 Transpiration provides the plant with water for photosynthesis, cell turgidity and mineral nutrients. The evaporation of water from leaves also produces a cooling effect for the plants.
- ☐ 12 Water vapour passes out of leaves through the stomata which open and close according to the condition of the guard cells. The stomata are open when the guard cells are turgid and closed when the cells are flaccid.
- ☐ 13 Xerophytes are plants adapted to survive in conditions that cause high transpiration rates and have adaptations such as deep roots, fleshy water-storing tissues, rolled leaves and sunken stomata.
- ☐ 14 Hydrophytes are adapted to living in water and have adaptations such as floating leaves with stomata on the upper surface only and long flexible leaf stalks.
- ☐ 15 In obtaining food, animals are adapted to use foraging strategies and search patterns that ensure that the energy used in searching for food is less than the energy gained. Behaviour must be organised to maximise energy gain.
- ☐ 16 The economics of foraging behaviour involve a net gain of energy by minimising the expenditure of energy in searching and the maximising the energy gained in the food obtained.
- ☐ 17 When resources are scarce competition between organisms occurs.
- ☐ 18 Interspecific competition occurs between different species whereas intraspecific competition occurs between organisms of the same species.
- ☐ 19 Some animals that live in large social groups have a dominance hierarchy in which the animals are in rank order from dominant to subordinate.
- ☐ 20 Some animals that live in large social groups use co-operative hunting techniques that benefit subordinate as well as dominant individuals since a subordinate animal may gain more food than by foraging on its own. Food sharing will occur as long as the reward for sharing exceeds that for foraging individually.
- ☐ 21 Co-operative hunting reduces each individual's energy expenditure; there is greater hunting success and larger prey can be caught.
- ☐ 22 Some animals show territorial behaviour in which the energy used in defending the territory is significantly less than the energy that can be obtained in searching for food in the absence of competition.
- ☐ 23 Plants are generally unable to move around their habitat and are said to be sessile. Animals on the other hand are mobile and move around their habitat in search of food.
- ☐ 24 Partly because of their inability to move, plants are often in competition with each other for light and soil nutrients.
- ☐ 25 Grazing by herbivores affects the diversity of species found in a habitat. Low and high levels of grazing tend to reduce diversity but moderate grazing increases species diversity.
- ☐ 26 Tall plants such as trees are sun plants because they always shadow smaller plants growing below them. Shade plants are adapted to grow beneath sun plants.
- ☐ 27 Shade plants reach their compensation point earlier in the day and at lower light intensities than sun plants. They are more efficient at carrying out photosynthesis at low light intensities.

Key Ideas *continued* ➢

Key Ideas continued

- ☐ 28 The compensation point is the light intensity at which a plant's rate of photosynthesis is the same as its rate of respiration and carbon dioxide uptake is the same as output.
- ☐ 29 Avoidance behaviour is used by animals to cope with dangers. An earthworm will withdraw into its burrow if stimulated by a potential predator and so avoid being eaten.
- ☐ 30 Habituation is a short-term modification of behaviour in which an organism learns to no longer respond to a harmless stimulus and so saves energy.
- ☐ 31 Learning is a long-term modification of response by animals.
- ☐ 32 Individual mechanisms for defence involve a single animal protecting itself from danger.
- ☐ 33 Social mechanisms of defence involve a group of animals acting in co-operation to discourage a predator.
- ☐ 34 Many plants have structural defence mechanisms such as stings, thorns and spines which discourage grazing animals.
- ☐ 35 Many plants are able to tolerate grazing by having low meristems that can rapidly regenerate. Others have deep root systems or underground stems that are not accessible to herbivores. These can help regenerate parts above ground that are damaged by grazing.

Topic Notes

Maintaining a water balance

Animals

In animals, water loss must be equalled by water gain.

Osmoregulation is the maintenance of water and solute concentration in tissues.

Freshwater fish tend to gain excess water by osmosis from their surrounding through their mouth lining and gills. Their kidneys are adapted to remove excess water in their urine. The kidneys have many large glomeruli and a high filtration rate, producing large volumes of dilute urine. Their gills contain chloride secretory cells that absorb salts by active transport from the surrounding water and pass these into the blood.

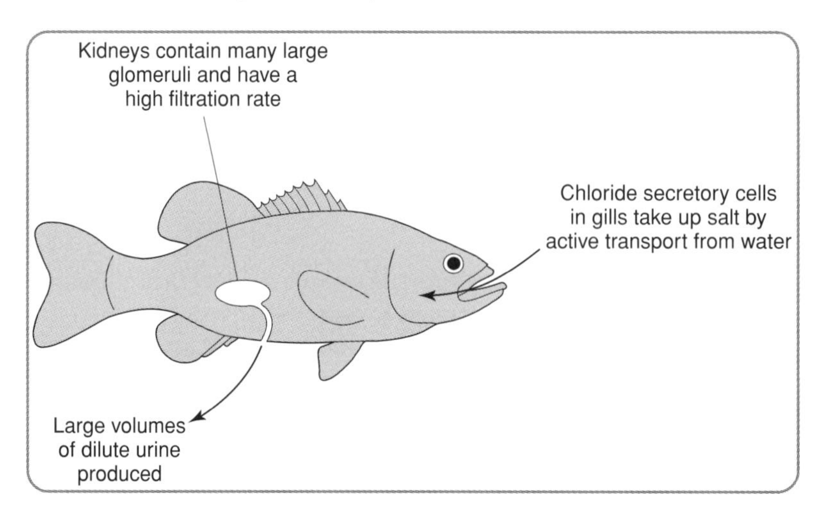

Figure 2.16 Adaptations of freshwater fish

Saltwater bony fish tend to lose excess water by osmosis to their surroundings through their mouth lining and gills. They drink salt water to replace the water being lost. Their kidneys are adapted to avoid further excess loss of water in their urine. The kidneys have few glomeruli and these are small in size giving a low filtration rate and producing small volumes of concentrated urine. Excess salt is removed from their bodies by the action of the chloride secretory cells in the gills. These cells excrete excess salt from the blood into the surrounding water by active transport.

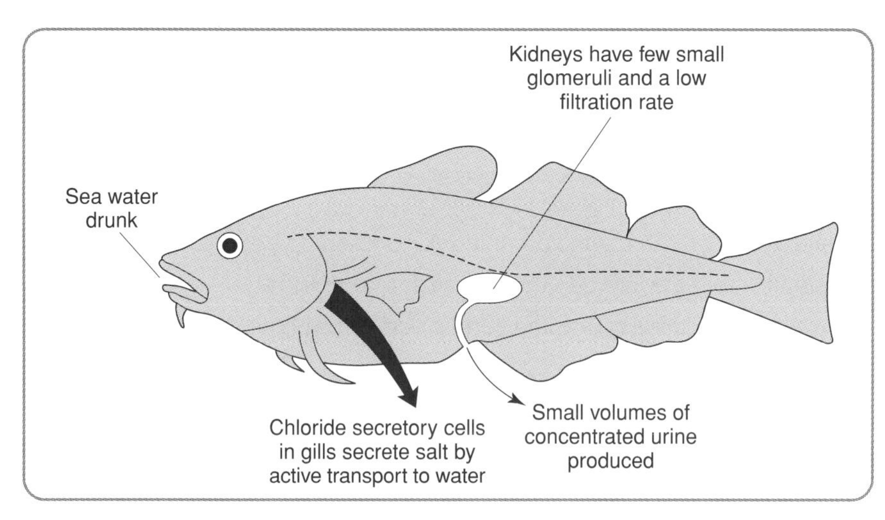

Figure 2.17 Adaptations of saltwater fish

Migratory species such as salmon and eel are able to alter their kidney function and reverse the action of their chloride secretory cells as they migrate between freshwater and saltwater. This seems to be regulated using a combination of hormones.

Desert mammals live in regions where very little free water exists. They have adaptations to conserve water. Adaptations can be physiological features of body functions or behavioural.

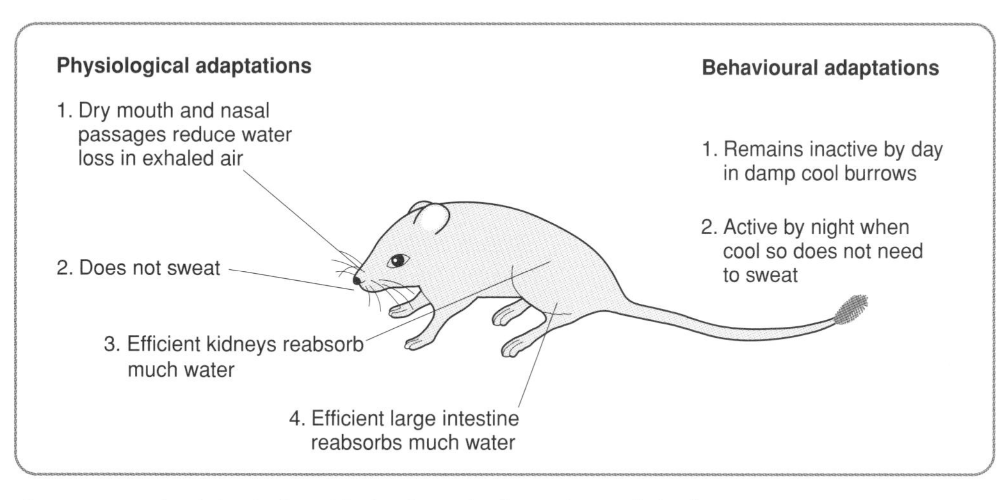

Figure 2.18 Physiological and behavioural adaptations of the kangaroo rat

Plants

Transpiration is the evaporation of water through the stomata on plant surfaces especially the underside of leaves.

Water moves through plants from the roots to the stems and out of the leaves in the transpiration stream.

Figure 2.19 shows the main tissues and processes involved in the transpiration stream. The processes are numbered 1–5.

1. The water enters the roots by osmosis into root hair cells that have a large surface area.
2. The water passes, again by osmosis, down the water concentration gradient through the cells of the cortex and into the xylem.
3. In the xylem, water molecules are held together by the force of cohesion. The force of adhesion attracts water molecules to the walls of the xylem. These two forces ensure that the threads of water that are present in xylem vessels do not break.
4. Water evaporates from the leaf cells into the air spaces of leaves.
5. The water vapour then diffuses into the atmosphere through the stomata.

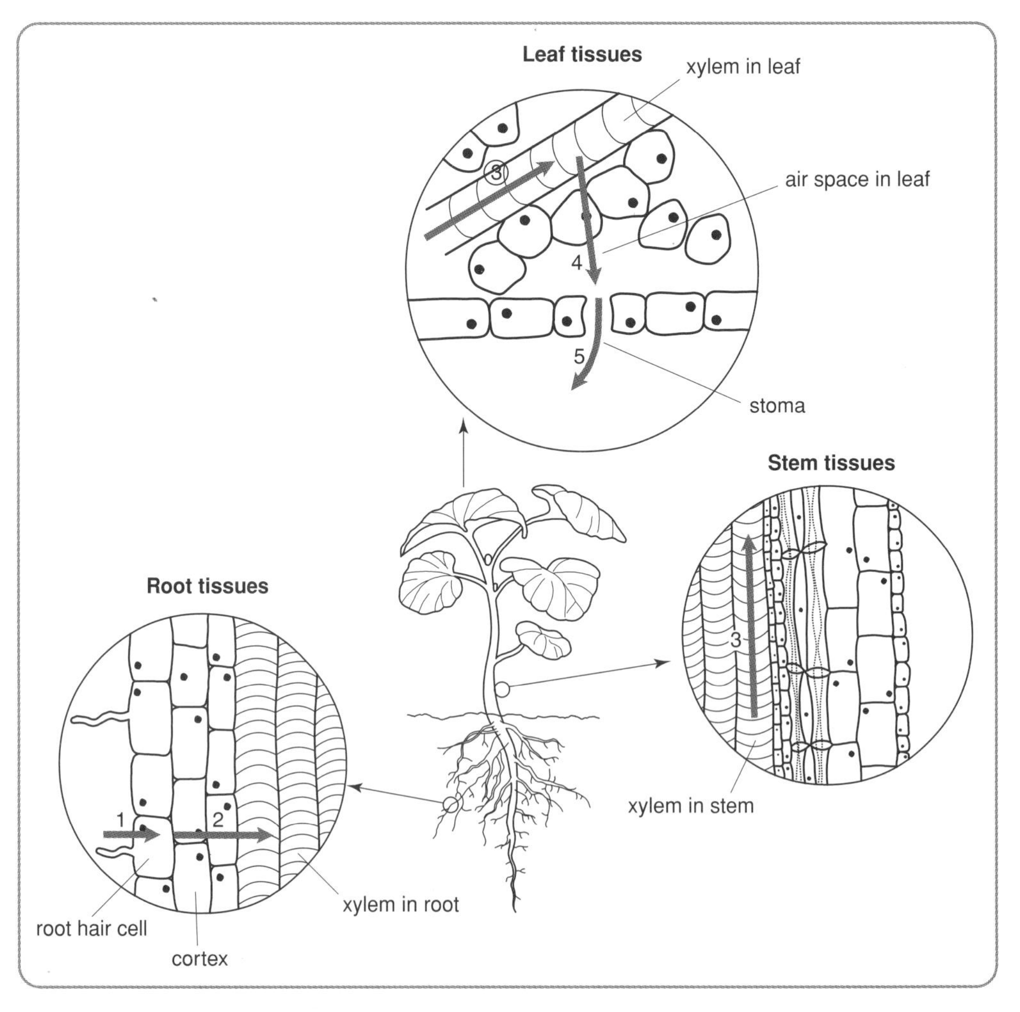

Figure 2.19 The transpiration stream

As well as supplying the leaves with water for turgidity and as a raw material in photosynthesis, the transpiration stream also supplies the plant with nutrient ions and has a cooling effect on the leaves.

The stomata open and close due to changes in turgor of the guard cells. When the guard cells are turgid, the pores are fully open. When the guard cells are flaccid the pores close.

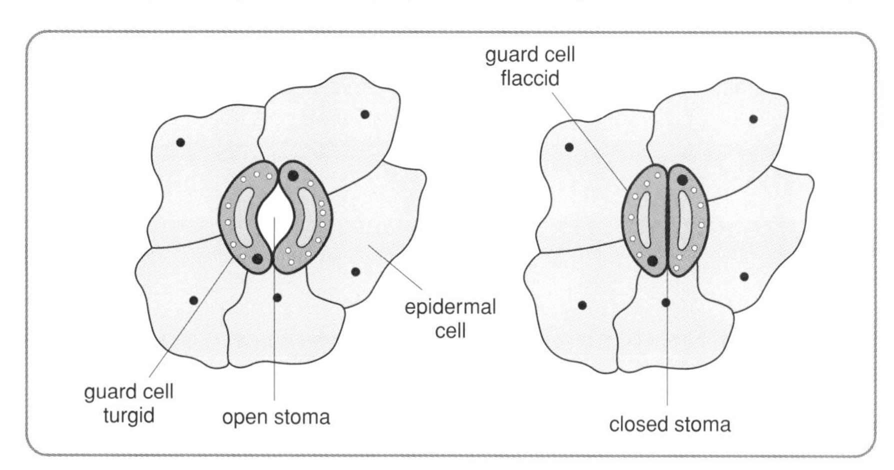

Figure 2.20 Two states of a stoma

Various factors and an explanation of their effects on the rate of transpiration are shown in the table.

Factor	Effect on transpiration	Explanation
Increased temperature	Increase	Increase in rate of evaporation
Increased humidity	Decrease	Decrease in rate of evaporation
Increased wind speed	Increase	Increase in rate of evaporation
Increased light	Increase	Stomata open to allow escape of water vapour by evaporation

Xerophytes are plants adapted to survive in conditions that would cause excessive transpiration. They can regulate their water balance in extremely hot, dry or windy conditions. Some xerophytes are adapted to survive temporary lack of water due to freezing soil conditions.

Hydrophytes are adapted to life in watery habitats such as flowing streams or ponds in which water levels can vary through the year.

Xerophyte adaptation	Explanation of effect
Thick waxy cuticle	Water cannot pass through so easily
Stomata sunk into pits	Increased humidity in stomatal pit decreases transpiration rate
Rolled leaves	Increased humidity and decreased air movements within the roll decrease transpiration rate
Hairy leaves	Increased humidity and decreased air movements within the hairy regions decrease transpiration rate

Hydrophyte adaptation	Explanation of effect
Stomata on upper leaf surface	Allow gas exchange with the air
Large air spaces in leaf	Allow the leave to float, keeping it in light and in contact with the air
Reduced central xylem	Allows flexibility of stem creating less resistance to flowing water
Long flexible leak stalks	Allow the leaf to move up and down with changing water levels

Top Tip

In your exam you could be asked to give an **explanation** of the effect of adaptations in xerophytes and hydrophytes. It is better to pick two or three from these two tables and concentrate on learning the **explanations** underlying each adaptation.

Obtaining food

Plants and animals have different strategies for obtaining food. These are based on the fact that plants make their own food by photosynthesis and are usually sessile (unable to move around) whereas animals have to search for food and are mobile.

Animals

The energy expended by animals in searching for food must be less than the energy obtained from the food. Foraging behaviour and search patterns must be organised to maximise **net** energy gain. Energy expended in foraging must be minimised and energy gained from prey items must be maximised. The graph (in Figure 2.21) shows a relationship between energy expended by a predator in foraging and energy gained as prey size increases.

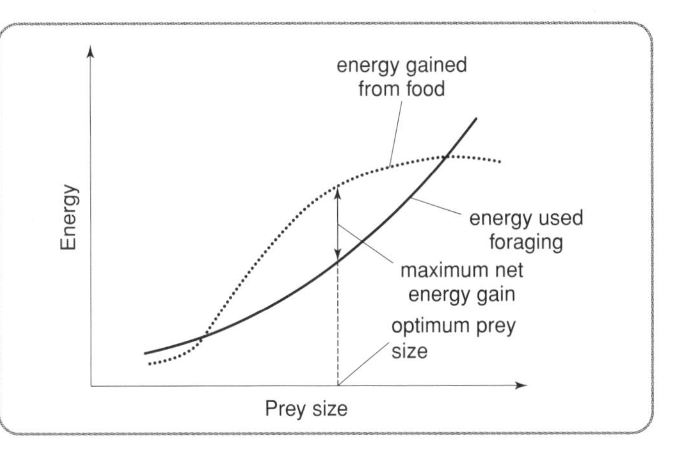

Figure 2.21 Energy expended in relation to prey size

Top Tip

In your exam you may see a reference to the economics of foraging – this simply means that energy gained from food must be greater than the energy expended in finding it! The term **net** gain of energy is useful as it refers to the **difference** between the **overall** gain and loss of energy in foraging.

When resources such as food are in short supply, animals come into competition with each other for the scarce resource. Intraspecific competition refers to the struggle between organisms of the same species for the resource. Interspecific competition refers to competition between organisms of different species for the same scarce resource.

Top Tip

Remember **inter**specific is **between different** species.
intraspecific is **within** the **same** species

Certain animal species live together in social groups. In some cases the social group shows a dominance hierarchy in which there is a rank order of individuals from dominant to subordinate. In feeding behaviour for example, dominant individuals feed before subordinate individuals. In some cases the social group uses co-operative hunting behaviour. This behaviour may benefit the subordinate individual as well as the dominant, as the subordinate animal may gain more food than by foraging on its own. Co-operative hunting reduces individual energy expenditure, increases hunting success and allows the catching of larger prey.

Certain animals mark and defend a territory. Occupation of a territory reduces competition for resources since the competitors are excluded. However, the energy expended in marking and defending the territory must be less than the energy gained from feeding exclusively within it.

Plants

In plants communities, competition is mainly for light and soil nutrient ions.

Grazing by herbivores affects the diversity of species within a plant community. Moderate grazing increases species diversity. Overgrazing reduces diversity and can lead to the extinction of less tolerant species. Under-grazing reduces diversity as strong competitors such as grasses dominate and block out light to other species.

In plant communities, sun plants such as forest trees are tall and so overshadow shade plants that grow beneath them. Both sun and shade plants are adapted to their position in the habitat. The compensation point is the light intensity at which a plant's rate of photosynthesis equals its rate of respiration, and its carbon dioxide uptake equals its output. Shade plants such as nettles reach their compensation point at lower light intensities and earlier in the day than sun plants such as sycamore trees.

Shade plants are more efficient at carrying out photosynthesis at low light intensities.

Coping with dangers

Animals

Avoidance behaviour is of survival value, keeping animals away from the danger of being eaten by potential predators. The marine fan worm *Sabella* lives in a tube from which it extends a fan of tentacles for feeding. If stimulated by a predator the worm reflexively withdraws its tentacles into the protective tube. This is avoidance behaviour. However, if the stimulus is repeated regularly and is found to be harmless, then the worm learns to no longer respond. This **short-term** modification of avoidance behaviour is called habituation. It helps the animal to balance energy as it prevents unnecessary response to harmless stimuli and allows the animal to continue feeding. The response must be short-term in case the next similar stimulus **is** harmful!

Long-term modification of response has to be learned. A young insect-eating bird may eat a ladybird. These insects are highly distasteful to the birds and the unpleasant experience may teach the individual young bird not to try these insects again. The warning red colour of the ladybird is easily recognised and so learning can be very rapid.

A variety of individual mechanisms for defence are found in the animal kingdom. The table gives a few examples.

Example	Explanation
Camouflage in peppered moths	Difficult for predators to spot
Mimicry in hoverflies	Resemble a stinging wasp so avoided
Possession of spines by hedgehogs	Passively discourages attack by predators
Spitting by fulmars	Actively discourages attack by predator

A variety of social mechanisms for defence are found in the animal kingdom. The table gives a few examples.

Example	Explanation
Crowding of gannet nests	Communal protection by neighbours thereby discouraging attack by predators
Flocking flight of starlings	Confusion of predator so difficult to pick out one prey item in flock
Flock feeding of wading birds	More eyes on look out for approaching predator so earlier alarm raised

Plants

Many plants have structural mechanisms for defence. These include sting, thorns and spines. These structures discourage grazing animals from attempting to eat them.

Top Tip

Remember that the structural mechanisms do not allow the plant to **tolerate** grazing. They simply **discourage** grazing.

Other plant species can tolerate grazing because, although they may be damaged, they can re-grow quickly. Some species have low meristems, often right at ground level, that are not reached by the grazer. These can quickly regenerate the rest of the plant after damage. Others have deep roots or underground stems that are out of the reach of the grazers and can regenerate upper parts quickly after damage.

Exam Questions

Marks

SBQ 1 The diagram shows a magnified region of a plant root.

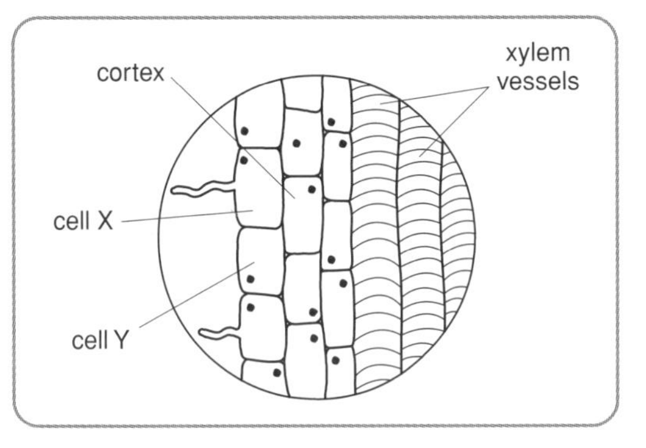

(a) Describe the process by which water moves into cell X. 1B

(b) Explain how cell X is better adapted to its function than cell Y. 1B

(c) Name the force which attracts water molecules to the xylem walls as they travel up the plant stem. 1C

(d) Describe the changes in the turgor of the guard cells when the stomata close. 1A

(e) The transpiration stream provides the plant cells with water for photosynthesis.
Give two other benefits to plants of the transpiration stream. 2C

SBQ 2 (a) The marine worm *Sabella* lives in a tube made from sand grains. During feeding it extends a fan of tentacles that trap small items of food.

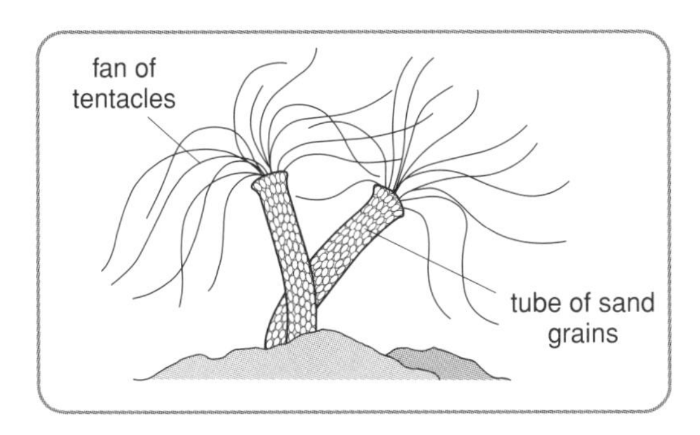

***Exam Questions** continued* ➢

Exam Questions continued

(i) Disturbance causes the withdrawal of the tentacles to the safety of the tube. They re-emerge minutes later.

1 Name the type of behaviour shown by the withdrawal of the tentacles. 1C

2 State the advantage to the worm of the withdrawal response. 1C

(ii) Repeated harmless disturbance causes the withdrawal response to cease.

1 Name the behaviour shown by this change in response. 1C

2 State the advantage to the worm in this change in response. 1C

(b) Nettles have stings on their leaf edges that discourage grazing by herbivores.
Give **two** other examples of structural defence mechanisms in plants. 1C

SCQ 1 Write notes on animals under the following headings:

(i) foraging behaviour 4

(ii) social mechanisms for obtaining food and for defence. 6

(10)

In question SCQ 2, ONE mark is available for coherence and ONE mark is available for relevance.

SCQ 2 Give an account of the osmotic problems of saltwater bony fish and describe how water balance is maintained in such fish. (10)

Answers on pages 140–2

Chapter 3

CONTROL AND REGULATION

Unit 3 is called 'Control and Regulation'. It comprises the following topic areas:

3.1 Control of Growth and Development

3.2 Physiological Homeostasis

3.3 Population Dynamics

3.1 Control of Growth and Development

Key Ideas

- ☐ 1 A meristem is a region of cells dividing by mitosis.
- ☐ 2 Plants have two types of meristems called apical meristems and lateral meristems.
- ☐ 3 The apical meristems are at the tips of the roots and shoots.
- ☐ 4 The activity of the apical meristems leads to growth in length.
- ☐ 5 The lateral meristems are found in the vascular bundles between the xylem and phloem and are made of a tissue called cambium.
- ☐ 6 The activity of the lateral meristems produces new cells and leads to an increase in the thickness of the stem. This process leads to the formation of annual rings.
- ☐ 7 Annual rings of trees are produced during the spring and summer and are made of xylem vessels.
- ☐ 8 In spring, the xylem vessels have thin walls and a wide diameter and in summer they have thicker walls but a narrow diameter.
- ☐ 9 The age of the tree can be calculated by counting the number of annual rings.
- ☐ 10 In animals, there are no meristems and growth takes place all over.
- ☐ 11 Regeneration is a process by which an organism replaces lost or damaged tissue.
- ☐ 12 Some organisms such as angiosperms show extensive powers of regeneration whereas others such as mammals have limited powers of regeneration.
- ☐ 13 Growth is the irreversible increase in the dry mass of an organism.
- ☐ 14 In an annual plant, a decrease in the dry mass occurs during germination when the seed is using its food store. It increases steadily in mass after photosynthesis starts. There can be a loss in mass due to seed dispersal.
- ☐ 15 A tree continues to grow and increase in mass throughout its life. A temporary decrease in mass occurs each autumn in species with leaf fall.
- ☐ 16 In humans, there is often a decrease in mass just after birth followed by an increase with two growth spurts, as a toddler and at puberty.
- ☐ 17 An insect such as a locust increases in length in a stepwise way due to the periodic shedding of the skin.
- ☐ 18 Scientists Jacob and Monod formed a hypothesis to explain gene action in bacteria with respect to lactose metabolism in *Escherichia coli* (*E. coli*).

Key Ideas *continued* ➢

Key Ideas continued

- ☐ 19 *E. coli* produces the enzyme needed to digest lactose only if lactose is present.
- ☐ 20 A regulator gene codes for the production of a repressor molecule.
- ☐ 21 The repressor protein blocks the operator which switches off the structural gene coding for the enzyme.
- ☐ 22 When lactose is present it binds to the repressor molecule and prevents it attaching to the operator. This then switches on the structural gene resulting in the production of the enzyme.
- ☐ 23 The enzyme continues to be produced until all the lactose is digested and so saves resources such as ATP and amino acids.
- ☐ 24 Since the presence of lactose is responsible for the switching on of the structural gene it is called the inducer molecule.
- ☐ 25 A metabolic pathway is a sequence of chemical reactions controlled by enzymes.
- ☐ 26 Each enzyme that controls a stage in a metabolic pathway is coded for by a specific gene.
- ☐ 27 If a gene which codes for an enzyme undergoes a mutation, either the enzyme will not be produced or it will be abnormal resulting in a metabolic block.
- ☐ 28 Phenylketonuria is an inherited disorder in humans caused by a failure to produce the specific enzyme needed to breakdown phenylalanine, an amino acid present in the diet.
- ☐ 29 The differentiation of cells depends upon which genes are switched on or off during their formation.
- ☐ 30 The pituitary gland is found at the base of the brain.
- ☐ 31 Two of the hormones produced by the pituitary gland, growth hormone (GH) and thyroid stimulating hormone (TSH), affect growth and development.
- ☐ 32 Growth hormone stimulates bone and muscle growth by increasing the rate of mitosis and increasing amino acid uptake and protein synthesis in cells.
- ☐ 33 The thyroid gland produces the hormone thyroxine in response to the TSH.
- ☐ 34 Thyroxine controls the metabolic rate and therefore affects growth.
- ☐ 35 Indole acetic acid (IAA) and gibberellic acid (GA) are plant growth substances.
- ☐ 36 IAA is produced in meristems at the root tips and shoot tips.
- ☐ 37 At the cell level, IAA stimulates mitosis, elongation and differentiation.
- ☐ 38 The apical bud produces high concentrations of IAA which inhibits the growth and development of the lateral buds and promotes the growth of the apical bud. This is called apical dominance.
- ☐ 39 Leaf abscission is caused by a decrease in the IAA concentration in the leaf stalk causing the leaf to fall off.
- ☐ 40 IAA causes fruit formation.
- ☐ 41 GA breaks bud dormancy in spring.
- ☐ 42 GA causes elongation of the internode cells in dwarf varieties of plants.
- ☐ 43 GA produced by seed embryos diffuses to the aleurone layer stimulating the production of amylase.
- ☐ 44 Amylase is responsible for the breakdown of starch to maltose which can then be used by the seed to provide energy for germination.
- ☐ 45 Plant growth substances have practical applications. IAA is used in herbicides and rooting powders.

Key Ideas continued ➢

Key Ideas continued

- ☐ 46 Macroelements are chemicals which are required for the normal healthy growth of plants. They include nitrogen, phosphorus, potassium and magnesium.
- ☐ 47 Nitrogen is required in amino acid and protein synthesis. Plants lacking nitrogen may have reduced shoot growth, chlorosis or pale green or yellow leaves, red leaf bases and a longer root system.
- ☐ 48 Phosphorus is required for compounds such as DNA, RNA, ATP, NADP, RuBP and GP. Plants lacking phosphorus may have overall growth reduced and red leaf bases.
- ☐ 49 Potassium is important in membrane transport and is also involved in protein synthesis and the activation of enzymes. Plants lacking potassium may have premature death of leaves and reduced growth.
- ☐ 50 Magnesium is required for chlorophyll formation. Plants lacking magnesium may have chlorosis or pale green or yellow leaves and reduced growth due to lack of photosynthesis
- ☐ 51 Iron is a component of the respiratory pigment haemoglobin responsible for the transport of oxygen as oxy-haemoglobin. It is also a component of many enzymes and hydrogen-carrying systems such as the cytochrome system. Deficiency in iron leads to anaemia.
- ☐ 52 Calcium is a mineral and is an essential component of shells, bones and teeth. It is also needed for blood clotting. Calcium deficiency in children results in a condition called rickets.
- ☐ 53 Lead inhibits the activity of enzymes in living organisms.
- ☐ 54 Vitamin D is required for the absorption of calcium from the intestines. It is also needed for the uptake of calcium into the bones and a deficiency in vitamin D leads to rickets in children.
- ☐ 55 Certain drugs taken during pregnancy can cross the placenta and affect the developing embryo.
- ☐ 56 Thalidomide is an anti-nausea drug which, when given to pregnant women to prevent morning sickness, caused limb deformities in their babies.
- ☐ 57 Alcohol taken during pregnancy can retard the growth and mental development of the baby.
- ☐ 58 Nicotine from cigarettes can retard the growth and mental development of the baby.
- ☐ 59 If grown in the absence of light a plant becomes etiolated. An etiolated plant has a long thin stem with long internodes and small yellow curled leaves.
- ☐ 60 Phototropism is the term used to describe the plant's growth movement in response to light from one side. The shoots grow towards the direction of the light.
- ☐ 61 The photoperiod is the number of hours of light in a day.
- ☐ 62 Photoperiodism is the response of an organism to changes in the photoperiod. Flowering in plants and the onset of courtship and breeding in birds and mammals are examples of photoperiodic responses.
- ☐ 63 Long day plants flower when the number of hours of light (photoperiod) is above a critical level. Short day plants flower when the number of hours of darkness is above a critical level.
- ☐ 64 The photoperiod affects the timing of breeding seasons in birds and mammals so that the young are born in the spring or summer when conditions are favourable and food is abundant.

Topic Notes

Growth differences between animals and plants

A plant meristem is a region of cells dividing by mitosis. The meristems produce new cells that then elongate to bring about growth. Plants have two types of meristems called apical meristems and lateral meristems.

Apical meristems are at the tips of roots and shoots. Activity of the apical meristems leads to growth in length. At the apex, the new cells are pushed back from the tips, vacuoles develop, water enters and the cells elongate then differentiate to become permanent tissue. Figure 3.1 shows these events in a root tip.

Lateral meristems, called cambium, are located in the vascular bundles between the xylem and phloem. Figure 3.2 shows the vascular bundles in a shoot and the position of the cambium.

Activity of cambium produces new cells leading to an increase in the thickness of the stem.

The annual rings that can be seen in a cut tree trunk are made of xylem vessels produced by cambium activity. Different sizes of xylem vessels are produced at different times of the year. In spring, the growing conditions are ideal and the xylem vessels produced are larger and have a wider diameter. In summer, the xylem vessels are smaller and have a smaller diameter. The spring and summer xylem tissue of a single year's growth gives a single ring. The age of the tree can be calculated by counting the number of annual rings.

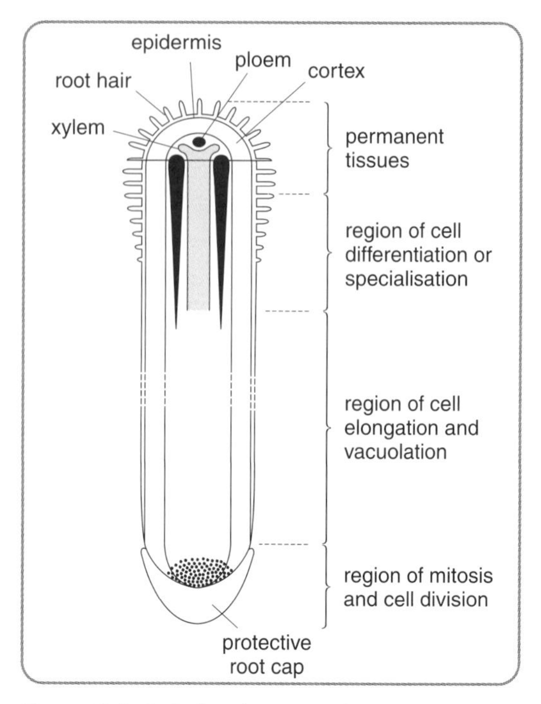

Figure 3.1 Cell development in a root tip

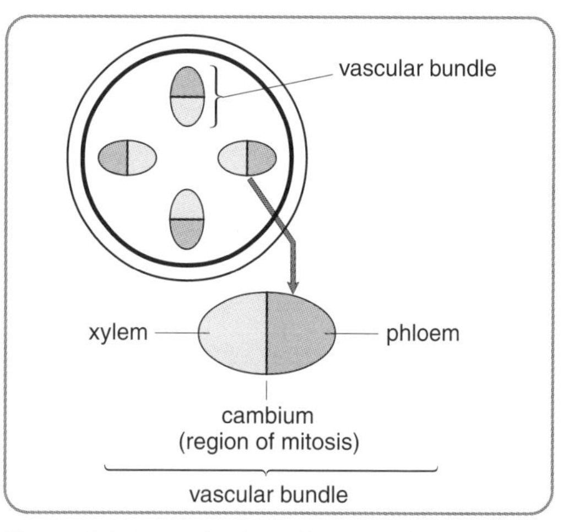

Figure 3.2 Vascular bundles in a shoot

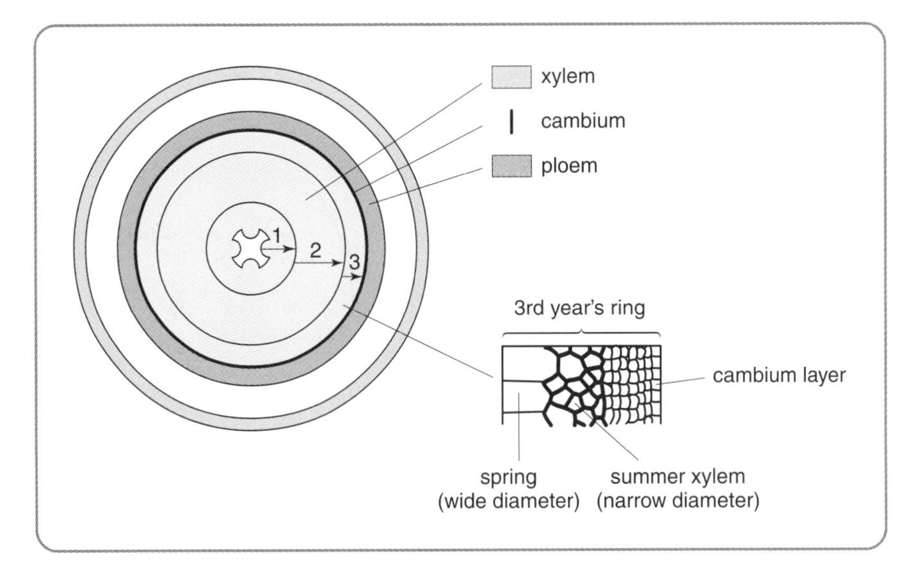

Figure 3.3 Annual rings in a three-year-old stem

In animals, there are no meristems and growth takes place all over.

Regeneration is a process by which organisms replace lost or damaged tissue.

Angiosperms (flowering plants) have extensive powers of regeneration. The process of regeneration involves specific genes being switched on in undifferentiated cells to allow them to differentiate into specialised tissue.

Mammals have limited powers of regeneration. Regeneration in mammals is limited to wound healing, repair of broken bones, and replacement of blood and of liver cells due to disease or damage. Mammals have few undifferentiated cells, most having already specialised to carry out a particular function.

Growth is the irreversible increase in the dry mass of an organism. Since measuring the dry mass kills the organism, growth is usually measured as an increase in the organism's fresh mass, height or length.

Top Tip

You are expected to explain why the increase in fresh mass is a less reliable measurement of growth than dry mass. This is because the fresh mass includes the water content that can vary from day to day.

The growth patterns of an annual plant, a tree, a human and a locust are shown in the following graphs. You should be able to recognise these growth patterns and explain their shapes.

In annual plants, the decrease in dry mass at the beginning (A) occurs during germination when the seed is using its food store. The plant starts to increase in mass when it starts to carry out photosynthesis. The loss in mass (B) is due to seed dispersal.

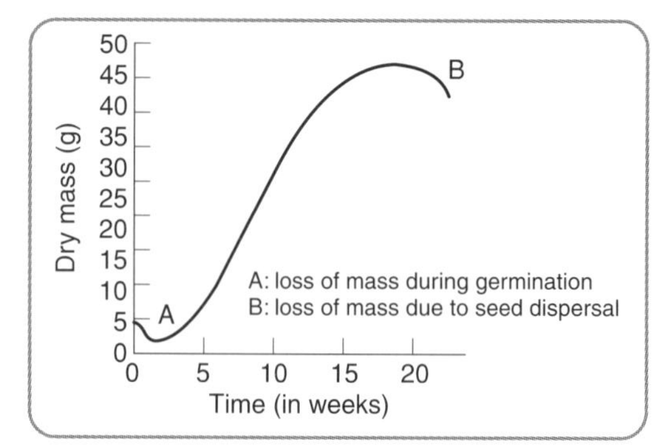

Figure 3.4 Growth pattern of an annual plant

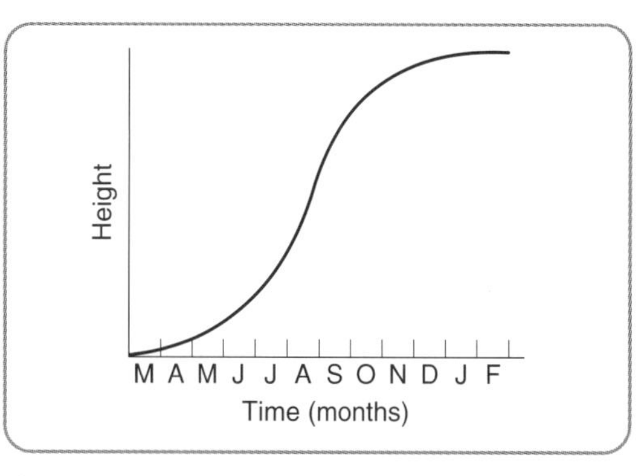

Figure 3.5 Growth pattern of a tree over one year

A tree continues to grow throughout its life. A decrease in mass occurs each autumn in species with leaf fall. The graph shows the growth pattern over one year with maximum rate of growth in the summer, when the rate of photosynthesis is highest.

In humans, there is often a decrease in mass just after birth. Humans have two fairly obvious growth spurts, as a toddler (A) and at puberty (B).

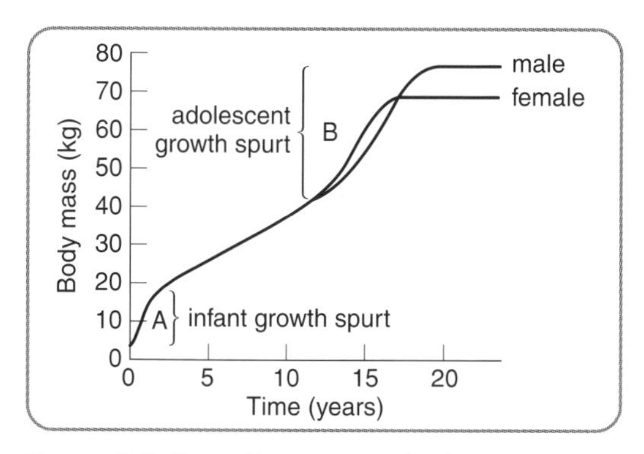

Figure 3.6 Growth pattern of a human

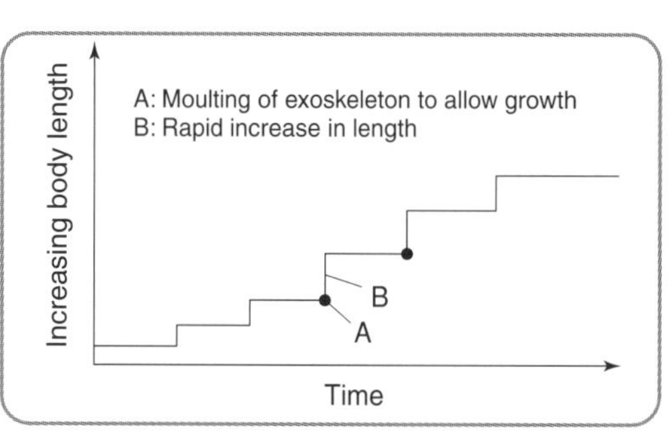

Figure 3.7 Growth pattern of a locust

This step-shaped growth curve is characteristic for an insect such a locust. The bottom point of each step (A) is when the locust moults and sheds its exoskeleton. This is immediately followed by a rapid increase in length (B) before its new exoskeleton hardens preventing further increase in length until the next month.

Top Tip

Do not use the terms outer **shell** or outer **coat** to describe the locust's exoskeleton.

Genetic control

The features of cells depend upon which of their genes are 'switched on' during their formation. Some genes that code for essential proteins such as the respiratory enzymes are 'switched on' in all cells. In the formation of red blood cells, the gene that codes for haemoglobin is 'switched on'. Genes not required by a cell remain 'switched off' which saves energy and resources.

Undifferentiated cells of a particular organism contain the same genetic information. As the organism develops its cells differentiate and become specialised to carry out different functions. The differentiation of the cells depends upon which genes are switched on or off during their formation.

The scientists Jacob and Monod suggested a hypothesis to explain the control of gene action in bacteria with respect to lactose metabolism in *Escherichia coli* (*E. coli*).

E. coli is able to produce the enzyme needed to digest the milk sugar lactose. The bacterium only produces the enzyme if lactose is present in its surroundings. The Jacob–Monod hypothesis attempts to explain how this control is achieved.

The hypothesis involves a group of three genes and their actions as shown in the table.

Gene	Action
regulator gene	codes for the production of the repressor molecule (protein) which can block the operator
operator	controls the switching on of the structural gene
structural gene	codes for the production of the **enzyme** needed to digest lactose

Top Tip

Remember **ROSE** – **R**egulator, **O**perator, **S**tructural, **E**nzyme.

The structural gene is only switched on in the presence of lactose, the inducer.

The diagrams in Figure 3.8 show the control system in the absence and presence of lactose.

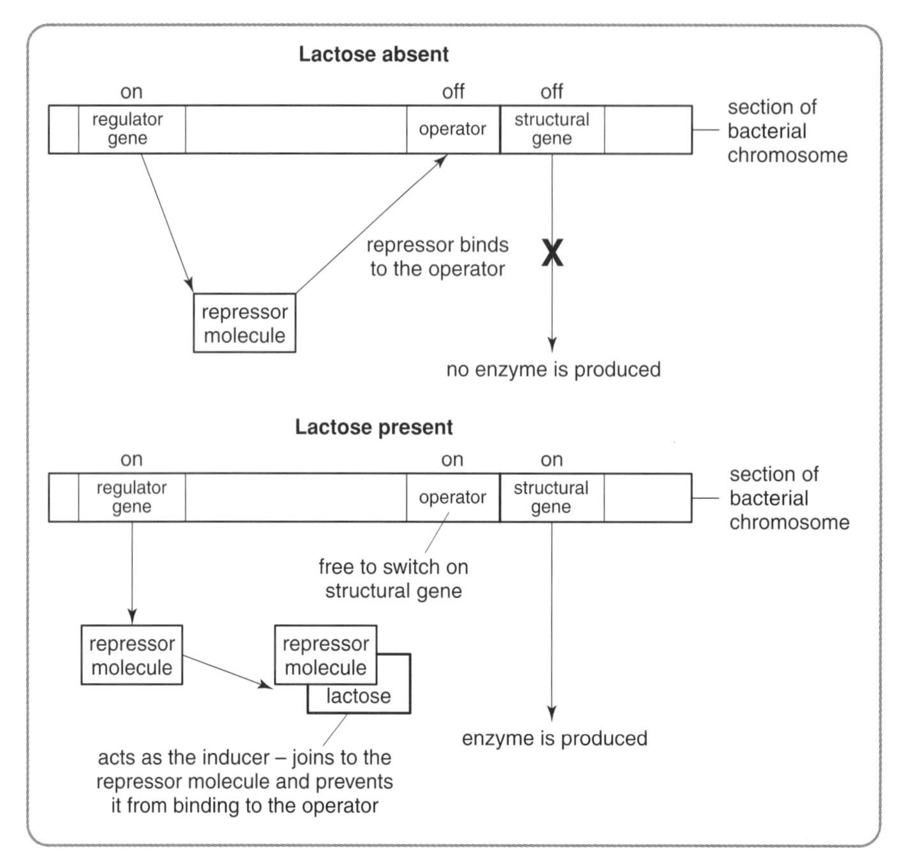

Figure 3.8 A genetic control system

In the absence of lactose the repressor molecule binds to the operator and prevents it switching on the structural gene. When lactose is present it binds to the repressor molecule and prevents it attaching to the operator. The operator is now free to switch on the structural gene that results in the production of the enzyme. The enzyme continues to be produced until all the lactose is digested.

Once the lactose is all digested the repressor molecule is released and can once more bind to the operator to switch off the structural gene.

This means that the enzyme is only produced when it is needed. This saves resources such as ATP and amino acids.

A metabolic pathway is a sequence of chemical reactions controlled by enzymes.

A specific gene codes for each enzyme that controls a stage in a metabolic pathway. The order of bases in the gene determines the order of the amino acids that determines the specific enzyme produced.

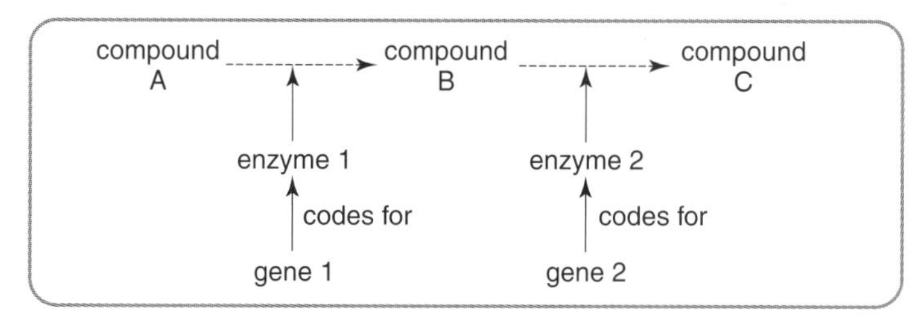

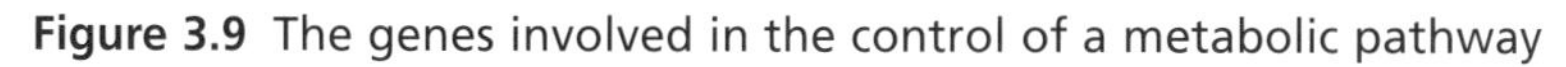

Figure 3.9 The genes involved in the control of a metabolic pathway

If a gene which codes for an enzyme undergoes a mutation – remember **DIGSI**? – then either the enzyme will not be produced or it will be abnormal or faulty resulting in a metabolic block as shown in Figure 3.10.

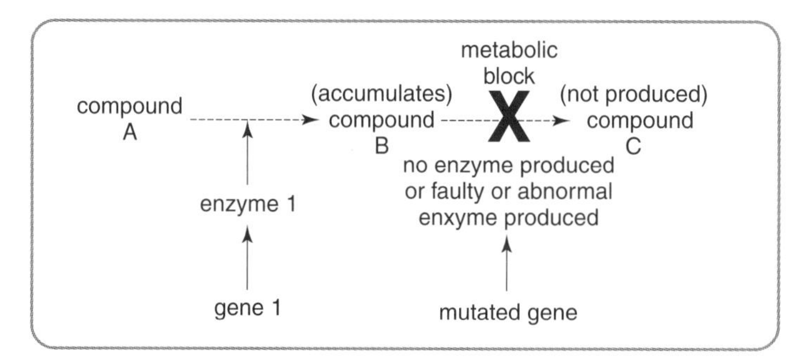

Figure 3.10 Mutated gene resulting in metabolic block

The part played by genes in controlling metabolic pathways can be shown in the case of phenylketonuria. Phenylketonuria is an inherited disorder in humans usually caused by a failure to produce the specific enzyme that breaks down phenylalanine, an amino acid in the protein we eat.

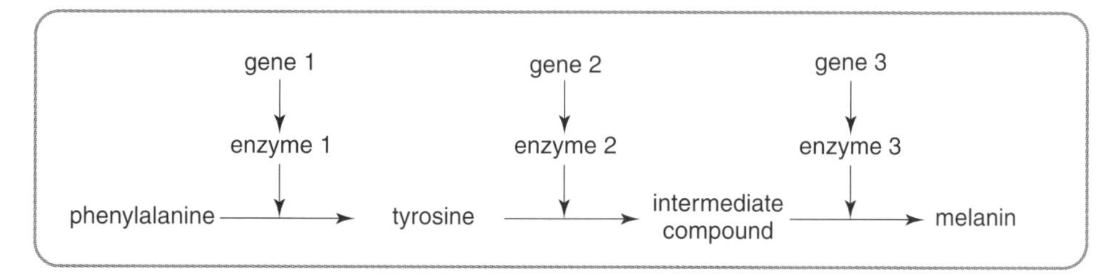

Figure 3.11 The normal metabolic pathway involving phenylalanine

A person who has inherited phenylketonuria has a mutation in gene 1 and, as a result, enzyme 1 is not produced. This causes a metabolic block. Phenylalanine undergoes a different metabolic pathway producing toxic compounds that affect developing brain tissue causing mental retardation (see Figure 3.12).

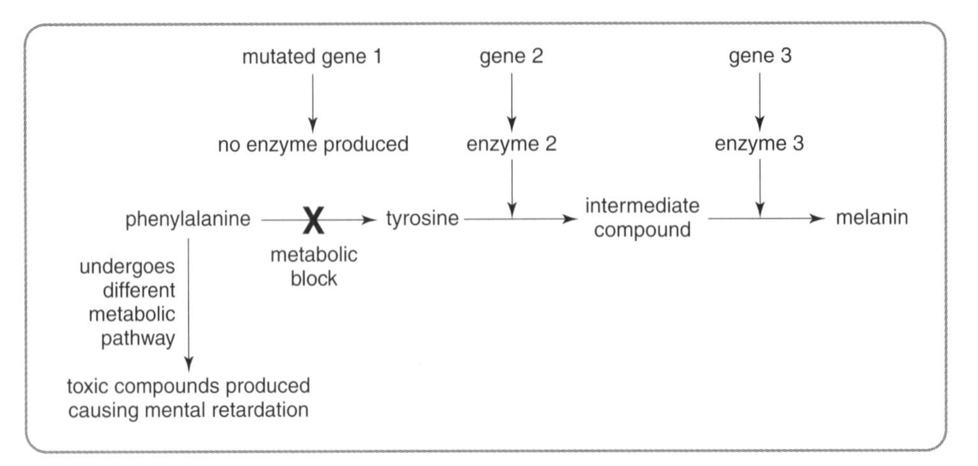

Figure 3.12 The effect of a metabolic block in phenylketonuria

Hormonal influences

Humans

The pituitary gland is found at the base of the brain. Two of the hormones produced by the pituitary gland affect growth and development. These are growth hormone (GH) and thyroid stimulating hormone (TSH).

Growth hormone stimulates bone and muscle growth by increasing the rate of mitosis and increasing amino acid uptake and protein synthesis in the bone and muscle cells.

Thyroid stimulating hormone (TSH) affects the thyroid gland in the neck. The thyroid gland produces the hormone thyroxine in response to the TSH. Thyroxine controls the metabolic rate and therefore affects growth.

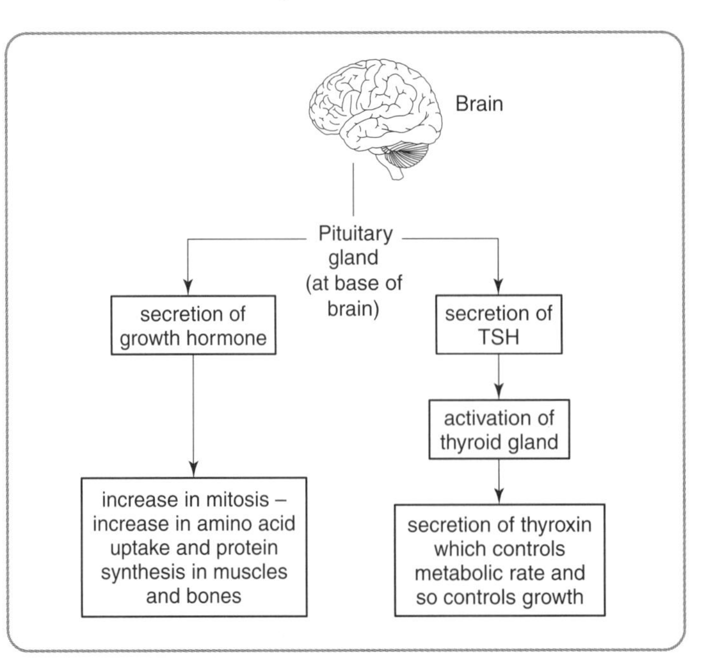

Figure 3.13 Effect of hormones produced by the pituitary gland

Plants

Indole acetic acid (IAA) and gibberellic acid (GA) are plant growth substances.

IAA is produced in meristems at root tips and shoot tips. The IAA diffuses from the tips to other parts of the plant. IAA affects individual plant cells as well as entire organs.

At the cell level, IAA stimulates mitosis, elongation and differentiation. IAA increases the elasticity of the cell wall that allows it to stretch and elongate when the vacuole forms.

Top Tip

Remember **MED-C**!

Mitosis
Elongation
Differentiation are at the
Cell level

IAA also affects plant organs such as the buds, leaves and fruits.

The apical bud produces high concentrations of IAA that inhibits the growth and development of the lateral buds and promotes the growth of the apical bud. This is called apical dominance. If the apical bud is removed the IAA concentration decreases and the lateral buds start to grow.

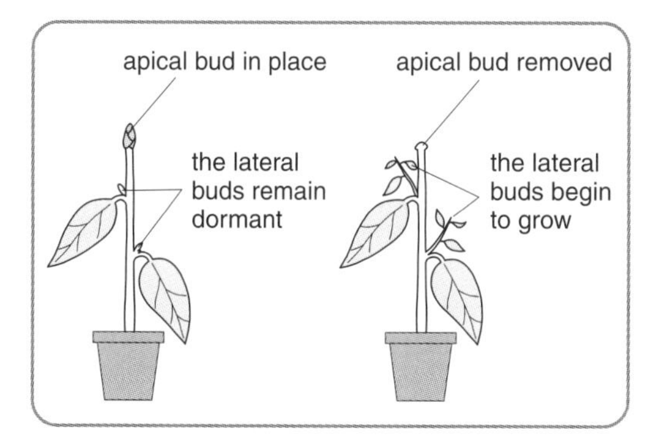

Figure 3.14 Effect of removing apical bud

Leaf fall is caused by a decrease in the IAA concentration in the leaf stalk. The decrease in the IAA causes an abscission layer to form and the leaf falls off.

After fertilisation the seeds produce IAA. This causes the ovary wall to develop into the wall of the fruit.

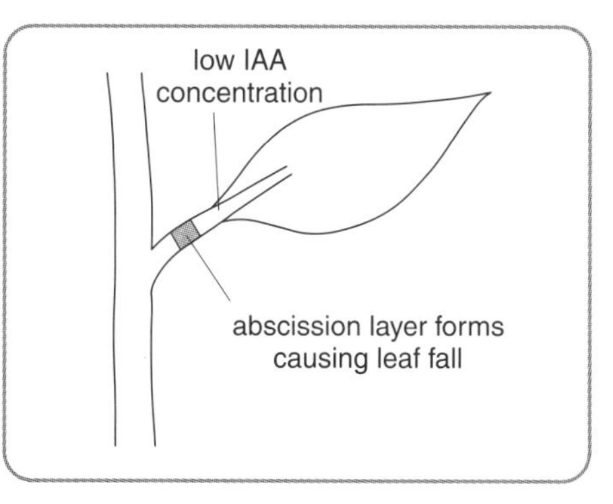

Figure 3.15 Process involved in leaf fall

Top Tip

Remember **OLAF**!

Organ level
Leaf abscission
Apical dominance
Fruit formation

GA breaks bud dormancy in spring.

GA causes elongation of the internode cells in the stems of dwarf varieties of plants. Dwarf varieties treated with GA grow to the same height as a normal variety.

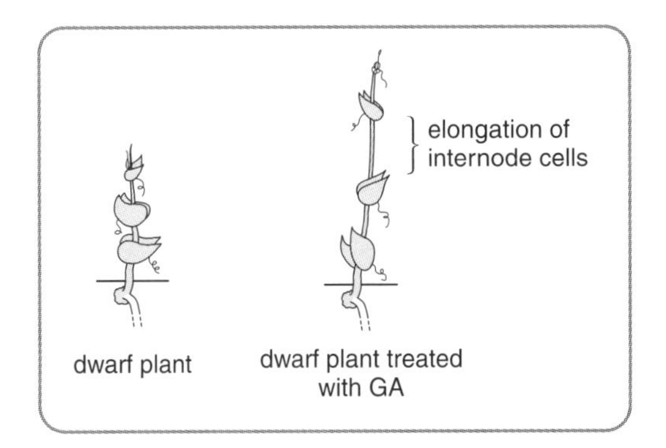

Figure 3.16 Effect of treating dwarf plant with GA

GA induces the production of the enzyme α-amylase in barley grains. GA breaks seed dormancy and causes seeds to germinate. The GA is produced by the embryo of the seed and then diffuses out to the aleurone layer. The GA stimulates the cells of the aleurone layer to produce the α-amylase. The α-amylase is responsible for the breakdown of starch to maltose that can then be used by the seed to provide energy for germination.

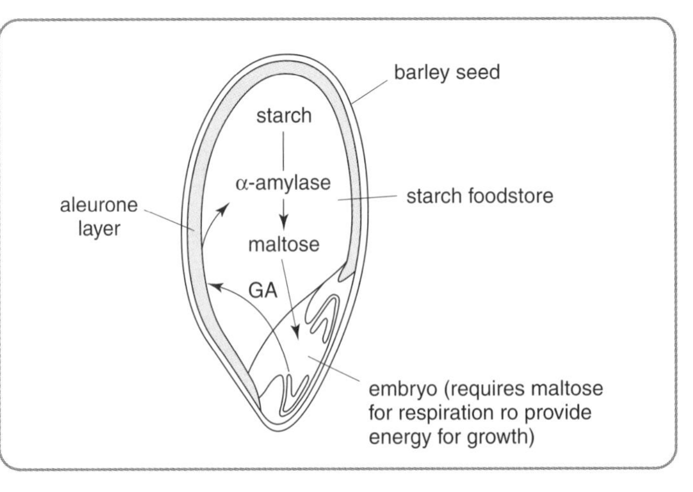

Figure 3.17 Processes in seed germination

Top Tips

SAM = **S**tarch ; **A**mylase ; **M**altose

GA – **AID** = **A**mylase ; **I**nternodes ; **D**ormancy

Growth substances have practical applications. IAA is used as a herbicide (selective weedkiller). These kill broad-leaved weeds by increasing the plants' metabolism and accelerating their growth making them use up all their food store. Narrow-leaved plants such as grass and cereal crops are not affected. IAA also stimulates cuttings to produce new roots.

Environmental influences

Macroelements are chemicals that are required for the normal healthy growth of plants. They include nitrogen, phosphorus, potassium and magnesium.

Nitrogen is required in amino acid and protein synthesis. Protein and therefore nitrogen is required for membrane structure and enzymes controlling all the cells' chemical reactions. Nitrogen is also required for the nucleic acids DNA and RNA that are essential for cell division and protein synthesis. It is also required for the pigment chlorophyll needed for photosynthesis.

Phosphorus is required for compounds such as DNA and ATP.

Potassium is important in membrane transport. It is also involved in protein synthesis and the activation of enzymes.

Magnesium is required for chlorophyll formation.

Top Tips

Nitrogen is needed for ami**N**o acid synthesis and protei**N**s

phospho**R**us = **R**ed leaf base

potassiu**M** = **M**embrane transport

The **M**i**N**us elements – chlorosis of the leaves occurs when the plant is **M**i**N**us **M**agnesium and/or **N**itrogen.

What You Should Know

Symptoms of deficiency of the macroelements.

- **Nitrogen** — reduced shoot growth; chlorosis or pale green or yellow leaves; red leaf base; longer root system
- **Phosphorus** — red leaf bases; overall growth reduced
- **Potassium** — premature death of leaves; reduced growth
- **Magnesium** — chlorosis or pale green or yellow leaves; reduced growth due to lack of photosynthesis

Iron is a component of the respiratory pigment haemoglobin, many enzymes and hydrogen-carrying systems such as the cytochrome system.
Deficiency in iron leads to anaemia.

Calcium is a mineral which is an essential component of shells, bones and teeth. It is also needed for blood clotting.
Calcium deficiency in children leads to rickets.

Lead inhibits the activity of enzymes in living organisms.

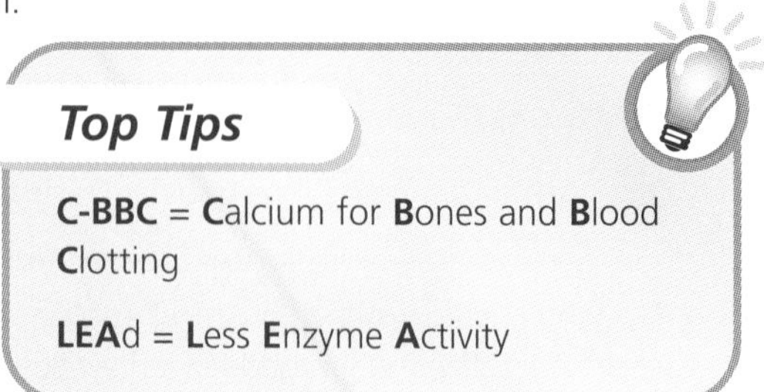

Top Tips

C-BBC = **C**alcium for **B**ones and **B**lood **C**lotting

LEAd = **L**ess **E**nzyme **A**ctivity

Vitamin D is required for the absorption of calcium from the intestines. A deficiency in vitamin D leads to rickets in children.

Certain drugs if taken during pregnancy can cross the placenta and affect the development of the fetus. You need to know the effect of thalidomide, alcohol and nicotine. The table shows these effects.

Drug	Description of effect on fetus
Thalidomide	Thalidomide is an anti-nausea drug which, when given to pregnant women to prevent morning sickness, caused limb deformities.
Alcohol	Alcohol taken during pregnancy can retard growth and mental development.
Nicotine	The nicotine from cigarettes can retard growth and mental development.

Top Tip

Pregnancy **ANT**i drug campaign i.e. **A** = alcohol ; **N** = nicotine ; **T** = thalidomide

In the absence of light a plant becomes etiolated. An etiolated plant has a long thin stem with long internodes and small yellow curled leaves. The survival value of this growth is that the plant increases its chances of reaching light needed for photosynthesis.

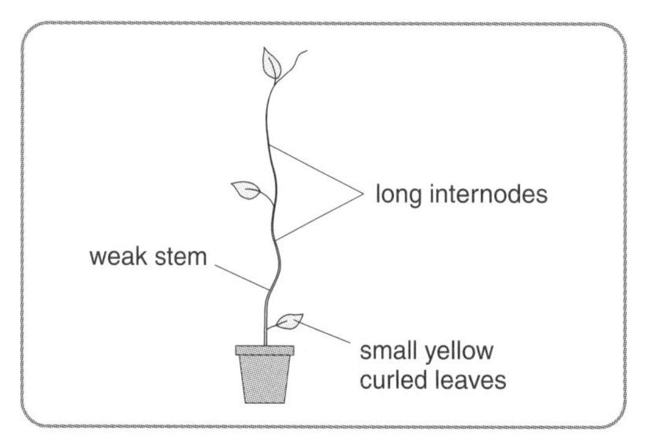

Figure 3.18 An etiolated plant

Phototropism is the term used to describe the plant's growth movement (tropism) in response to light (photo). If light comes from one side only, the shoots grow towards the direction of the light. The benefit to the plant is that it obtains light for its leaves to carry out photosynthesis.

The photoperiod is the number of hours of light in a day. Photoperiodism is the response of an organism to changes in the photoperiod.

Top Tip

DO NOT use the term day-length since the length of a day is **always** 24 hours and this includes hours of light and dark.

Flowering in plants and the onset of courtship and breeding in birds and mammals are examples of photoperiodism.

Long day plants flower when the number of hours of light (photoperiod) is above a critical level. Long day plants tend to flower in spring through to midsummer.

Short day plants flower when the number of hours of darkness is above a critical level. Short day plants tend to flower between late summer and autumn.

The photoperiod affects the timing of breeding seasons by controlling the production of sex hormones and gametes. The photoperiod triggers the onset of gamete production, courtship and breeding. The photoperiod required depends on the gestation period.

The long day breeders that include small mammals, such as rabbits, and birds, which breed in the spring as the photoperiod increases. They have a short gestation period (length of pregnancy) or period of incubation (for birds) and this ensures that the young are born in the spring or summer when conditions are favourable and food is abundant.

Large mammals such as sheep and deer are short day breeders. They breed in autumn as the photoperiod decreases. They have a longer gestation period and so this ensures that the young are born in the following spring when the conditions are favourable and food is abundant. This gives them a long period of growth before winter.

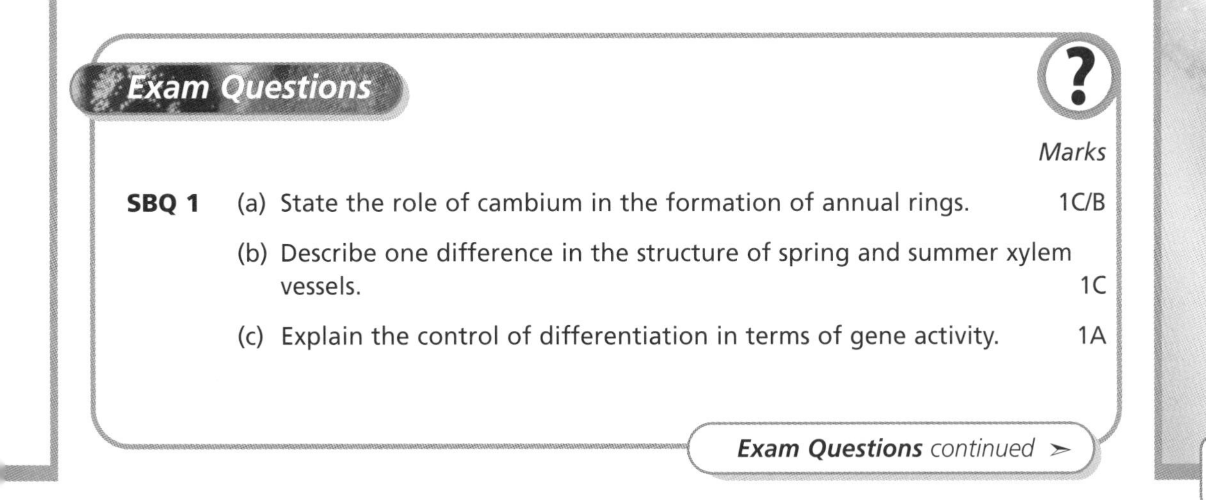

Exam Questions

Marks

SBQ 1 (a) State the role of cambium in the formation of annual rings. 1C/B

(b) Describe one difference in the structure of spring and summer xylem vessels. 1C

(c) Explain the control of differentiation in terms of gene activity. 1A

***Exam Questions** continued* ➢

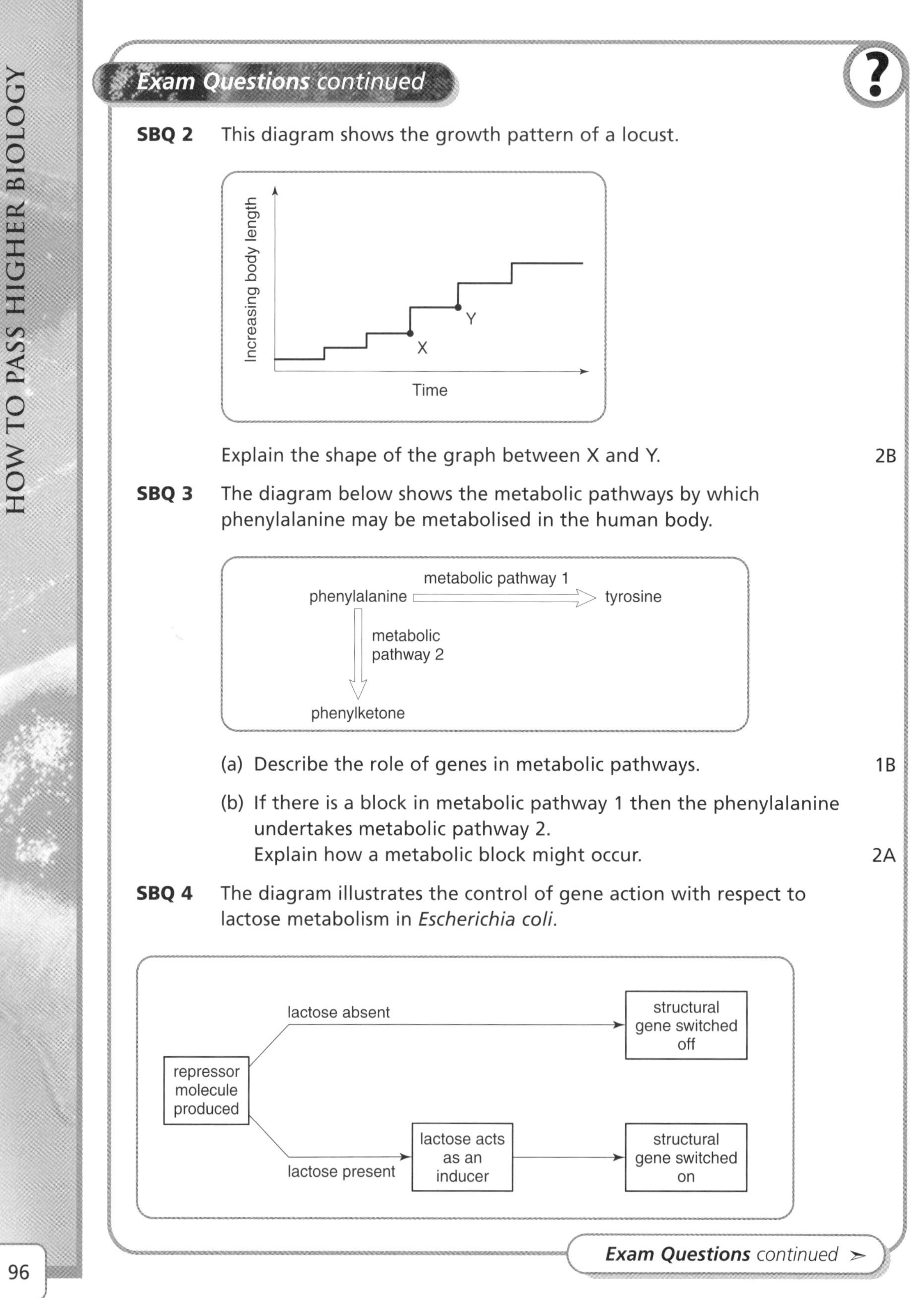

Exam Questions continued

SBQ 2 This diagram shows the growth pattern of a locust.

Explain the shape of the graph between X and Y. 2B

SBQ 3 The diagram below shows the metabolic pathways by which phenylalanine may be metabolised in the human body.

(a) Describe the role of genes in metabolic pathways. 1B

(b) If there is a block in metabolic pathway 1 then the phenylalanine undertakes metabolic pathway 2.
Explain how a metabolic block might occur. 2A

SBQ 4 The diagram illustrates the control of gene action with respect to lactose metabolism in *Escherichia coli*.

Exam Questions *continued* ➢

Exam Questions continued

(a) Explain why the structural gene remains switched off in the absence of lactose. 1B

(b) Explain how lactose acts as an inducer. 1A

(c) Give an advantage to *E. coli* of having this type of genetic control system. 1C

SBQ 5 (a) Complete the table below by inserting a tick in the appropriate box to show the role of the plant growth substances indole acetic acid (IAA) and gibberellic acid (GA) with respect to the processes listed.

Process	IAA	GA	Both
Stimulates amylase production in barley grains			
Inhibits leaf abscission			
Promotes fruit formation			
Stimulates stem elongation			
Breaks bud dormancy			
Causes apical dominance			

3B

(b) Give a practical application of plant growth substances. 1C

(c) Explain the importance of nitrate for the growth of barley plants. 1B

SCQ 1 Write notes on each of the following:

(i) the importance of nitrogen and magnesium in plant growth and development and symptoms of their deficiency 6

(ii) the importance of vitamin D and iron in humans. 4

(10)

In question SCQ 2, ONE mark is available for coherence and ONE mark is available for relevance.

SCQ 2 Give an account of the effect of light on shoot growth and development, and on the timing of flowering in plants and breeding in animals. (10)

Answers on page 142–5

3.2 Physiological Homeostasis

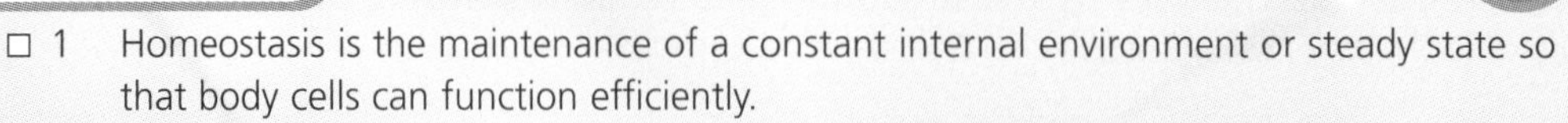

Key Ideas

- ☐ 1 Homeostasis is the maintenance of a constant internal environment or steady state so that body cells can function efficiently.
- ☐ 2 The steady state requires maintenance of conditions within tolerable limits.
- ☐ 3 Three examples of homeostatic mechanisms are the control of water concentration in the blood, the control of glucose concentration in the blood, and the control of body temperature.
- ☐ 4 Negative feedback allows homeostasis to be achieved.
- ☐ 5 Any change in the level of a factor is detected by receptors that switch on a corrective mechanism to restore the conditions back to normal. The corrective mechanism is switched off when the normal level is achieved.
- ☐ 6 Negative feedback systems have monitoring centres with receptor cells, a system for sending messages and effectors.
- ☐ 7 The receptors are special cells that constantly monitor internal factors and detect any changes in their level. Changes trigger a corrective mechanism and messages are sent to the effectors.
- ☐ 8 These messages can either be hormones in the blood or nerve impulses.
- ☐ 9 The effectors are the parts of the body that respond to the message. The effectors then counteract the change and return the conditions to normal.
- ☐ 10 Osmoregulation is the name given to the control of water and cell chemical concentration.
- ☐ 11 The water concentration of the blood is monitored by the hypothalamus in the brain.
- ☐ 12 Osmoreceptors in the hypothalamus detect changes in the water concentration of the blood flowing through them and send nerve impulses to the pituitary gland.
- ☐ 13 The pituitary gland releases more or less of the hormone ADH that travels in the bloodstream to the effectors, the kidney tubules.
- ☐ 14 ADH increases the permeability of the kidney tubules to water.
- ☐ 15 A decrease in the water concentration of the blood results in more ADH release and more reabsorption of water to the blood. A small volume of concentrated urine is produced.
- ☐ 16 An increase in the water concentration of the blood results in less ADH release and less reabsorption of water to the blood. A large volume of dilute urine is produced.
- ☐ 17 The pancreas monitors the blood glucose concentration in the blood.
- ☐ 18 The liver acts as a reservoir of stored carbohydrate, storing excess glucose as the carbohydrate glycogen. Since glycogen is insoluble it does not have an osmotic effect.
- ☐ 19 An increase in blood glucose concentration is detected by glucose receptor cells in the pancreas. This causes the pancreas to produce the hormone insulin.
- ☐ 20 Insulin travels in the blood to the liver, the effector. The insulin stimulates the liver to take up glucose and store it as glycogen reducing the blood glucose concentration.
- ☐ 21 A decrease in blood glucose concentration is detected by a group of receptor cells in the pancreas. This causes the pancreas to produce the hormone glucagon.
- ☐ 22 Glucagon travels in the blood to the liver, the effector. The glucagon causes the liver cells to convert glycogen to glucose increasing the blood glucose concentration.

***Key Ideas** continued* ➢

Key Ideas continued

- ☐ 23 In a situation of stress or danger the body can prepare itself for increased activity by releasing more glucose from glycogen.
- ☐ 24 The brain causes the adrenal glands above the kidneys to produce a hormone called adrenaline which converts glycogen to glucose and creates a higher than normal blood glucose concentration.
- ☐ 25 The control of body temperature is called thermoregulation.
- ☐ 26 Animals can be either ectotherms or endotherms.
- ☐ 27 Ectotherms are animals that do not have a homeostatic mechanism to regulate their body temperature.
- ☐ 28 Ectotherms' body temperature varies with the external environment and body heat is mostly gained from their surroundings.
- ☐ 29 Endotherms (mammals and birds) have a homeostatic mechanism to regulate their body temperature.
- ☐ 30 Endotherms are able to maintain a constant body temperature independent of the external environment and obtain most of their body heat from their own metabolism.
- ☐ 31 All the chemical reactions that take place in living cells are controlled by enzymes.
- ☐ 32 Enzymes have an optimum temperature at which they work best. Animals that can maintain the optimum temperature for enzyme activity have a high metabolic rate and can function more efficiently.
- ☐ 33 The hypothalamus is the temperature-monitoring centre. It contains thermoreceptors which detect changes in blood temperature.
- ☐ 34 The hypothalamus sends out nerve impulses to the effectors, skin and body muscles.
- ☐ 35 Increase in the body temperature results in an increase in sweat production from the skin. Heat from the body evaporates water in the sweat and lowers the body temperature.
- ☐ 36 Overheating results in vasodilation. The skin arterioles become dilated allowing a larger volume of blood to flow through the capillaries on the skin surface. The blood loses heat by radiation.
- ☐ 37 Decrease in the body temperature results in a decrease in sweat production and so less heat is used to evaporate the water in the sweat.
- ☐ 38 Overcooling results in vasoconstriction. The skin arterioles become constricted reducing the volume of blood that flows through the capillaries on the skin surface. Less heat is lost by radiation.
- ☐ 39 The hair erector muscles in skin contract to raise hairs. The air trapped by the hairs provides insulation and reduces heat loss.
- ☐ 40 In cool conditions, shivering of the muscles generates heat. There is an increase in the metabolic rate that increases heat production.

Topic Notes

Introduction

Physiology is the study of functions and processes in the body. Homeostasis is the maintenance of a constant internal environment or steady state, so that the body cells can function efficiently. The steady state requires maintenance of factors within tolerable limits. The factors you need to know for your exam are water concentration in the blood, glucose concentration in the blood and body temperature.

Negative feedback is the way in which homeostasis is achieved.

Top Tip

Learn this definition of negative feedback control.

Any change away from the optimum is detected by receptors that switch on a corrective mechanism to restore the conditions back to normal. The corrective mechanism is then switched off.

Negative feedback systems have monitoring centres with receptor cells, a system for sending messages and effectors. The receptors are special cells that constantly monitor the internal environment and detect any changes in the condition. If they detect a change away from the norm a corrective mechanism is switched on and messages are sent to the effectors. The messages can be either hormones in the blood or nerve impulses. The effectors are the parts of the body that respond to the message. The effectors then counteract the change and return the conditions to their optimum or norm.

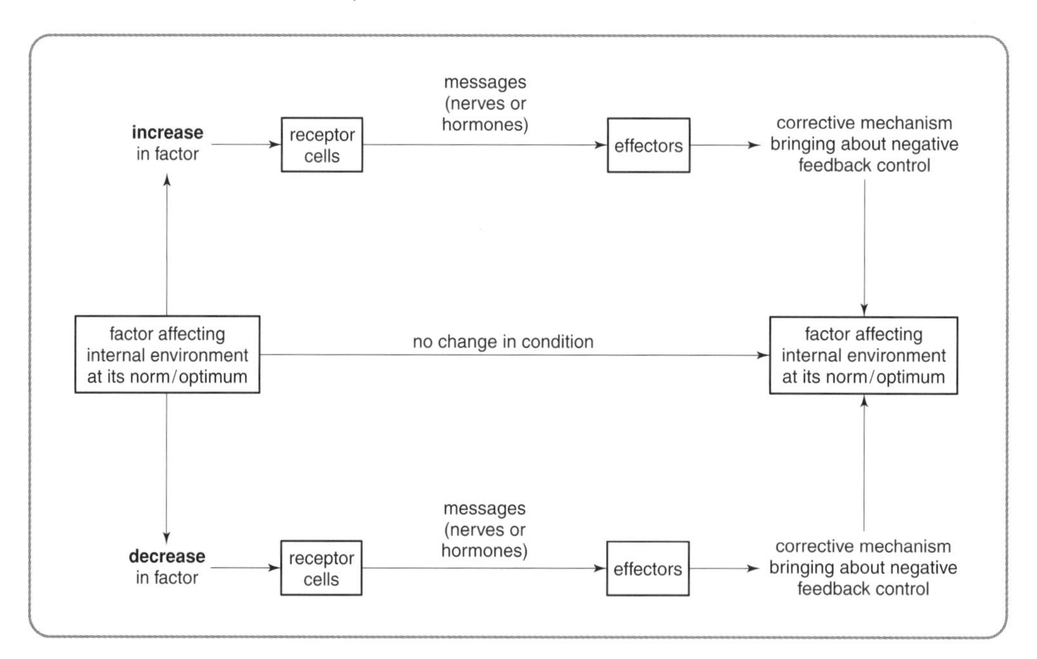

Figure 3.19 A general summary of negative feedback control

Control of water concentration

In Unit 1 you learned that animal cells such as red blood cells burst in hypotonic solutions and shrink in hypertonic solutions. Since cells can be damaged by an osmotic imbalance it is important to have a mechanism to control the water and chemical concentration in the blood. Osmoregulation is the name given to the control of water and salt concentration.

Top Tip

It is worth revising the structure and function of the kidney tubule from Standard Grade or Intermediate 2.

A monitoring centre in the brain, called the hypothalamus, detects the water concentration of the blood.

Osmoreceptors in the hypothalamus detect changes in the water concentration of the blood flowing through them. Changes in the water concentration detected by the osmoreceptors cause the pituitary gland to release more or less of the hormone ADH which travels in the bloodstream to the effectors, the kidney tubules.

Increased sweating, reduced fluid intake or eating salty food can cause a decrease in the water concentration of the blood. A decrease in the water concentration detected by the osmoreceptors in the hypothalamus causes the pituitary gland to release MORE ADH. MORE ADH makes the kidney tubules MORE PERMEABLE and so MORE water is REABSORBED back into the bloodstream. LESS URINE is produced although it is concentrated. This corrective mechanism returns the water concentration in the blood to normal.

Top Tip

For a decrease in water concentration – **MORE**; **MORE**; **MORE**; **LESS**.

MORE ADH therefore tubule **MORE** PERMEABLE therefore **MORE** WATER REABSORBED therefore **LESS** URINE produced.

Drinking a large volume of water can cause an increase in the water concentration of the blood. An increase in the water concentration detected by the osmoreceptors in the hypothalamus causes the pituitary gland to produce LESS ADH. LESS ADH makes the kidney tubules LESS PERMEABLE and so LESS WATER is REABSORBED back into the bloodstream. MORE URINE is produced although it is dilute. This corrective mechanism returns the water concentration in the blood to normal.

Top Tip

For an increase in water concentration – **LESS**; **LESS**; **LESS**; **MORE**.

LESS ADH therefore tubule **LESS** PERMEABLE therefore **LESS** WATER REABSORBED therefore **MORE** URINE produced.

Figure 3.20 summarises the control of water concentration in the blood.

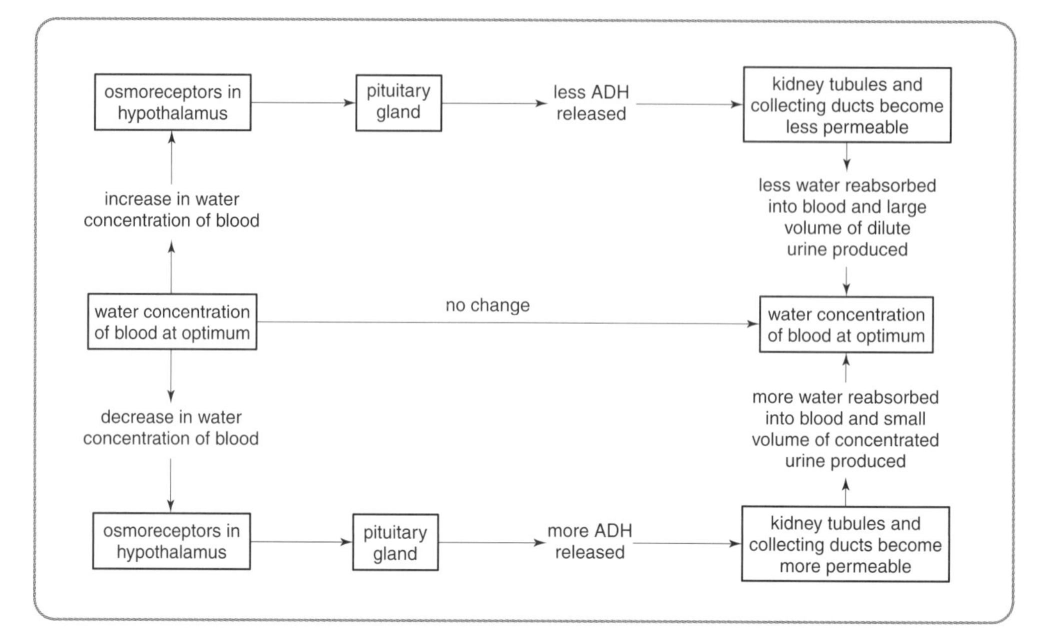

Figure 3.20 A summary of the control of water concentration in the blood

Control of glucose concentration

Living cells need a constant supply of energy in the form of ATP which they obtain by the oxidation of glucose during aerobic respiration. A homeostatic mechanism ensures that there is a steady glucose concentration in the blood.

Glucose receptors in the pancreas monitor the glucose concentration in the blood. The liver acts as a reservoir of carbohydrate and stores extra glucose as the insoluble carbohydrate glycogen.

Carbohydrate intake by eating increases the blood glucose concentration.
A rise in blood glucose concentration is detected by a group of receptor cells in the pancreas. This causes the pancreas to produce the hormone called insulin.
Insulin travels in the blood to the liver, the effector. The insulin stimulates the liver to take up glucose and store it as glycogen thereby reducing the blood glucose concentration.
This corrective mechanism returns the blood glucose concentration to normal.
Failure to produce enough insulin can lead to diabetes.

Missing a meal or taking a lot of exercise can result in a decrease in the blood glucose concentration. A decrease in blood glucose concentration is detected by a group of receptor cells in the pancreas. This causes the pancreas to produce the hormone glucagon. Glucagon travels in the blood to the liver, the effector. The glucagon causes the liver cells to convert glycogen to glucose thereby increasing the blood glucose concentration.
This corrective mechanism returns the blood glucose concentration to normal.

Top Tip

GLUCAGON

needed when

GLUCOSE is GONE

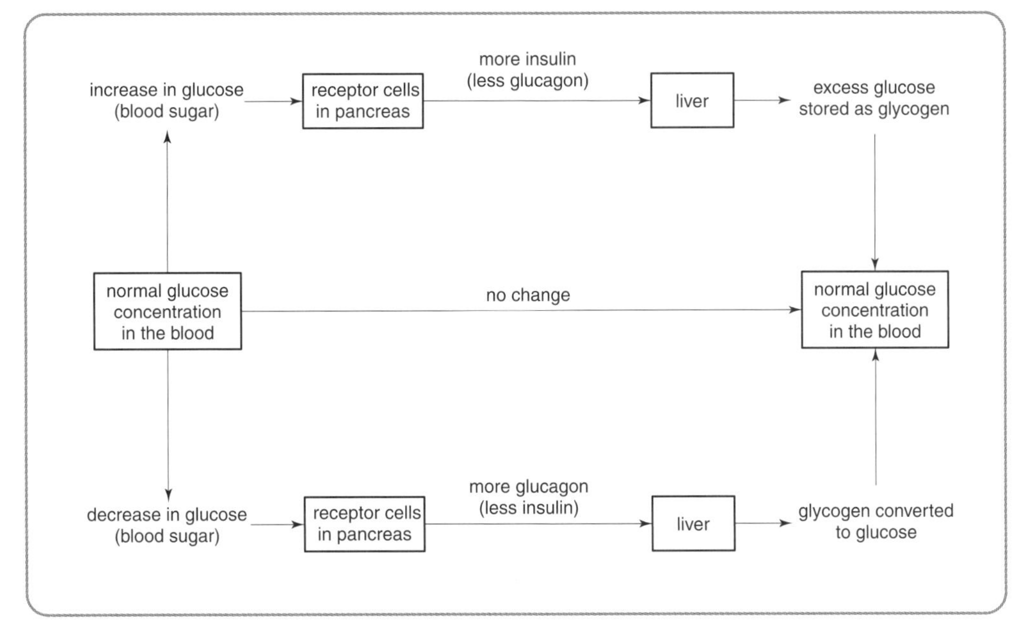

Figure 3.21 A summary of the control of glucose concentration in the blood

In a situation of stress or danger the body can prepare itself for increased activity by releasing more of the liver's stored glycogen as glucose. The brain causes the adrenal glands above the kidneys to produce a hormone called adrenaline which converts glycogen to glucose and creates a higher than normal blood glucose concentration.

Top Tips

A useful rhyme...

Low blood sugar – glucose gone
What you need is glucagon.
To turn glucose into glycogen
What you need is insulin.

This is one occasion when spelling is very important. You must not confuse **glycogen** and **glucagon**.

Control of body temeprature

The control of body temperature is called thermoregulation.

Animals can be either ectotherms or endotherms.

Ectotherms do not have a homeostatic mechanism to regulate their body temperature. Their body temperature varies with the external environment and they obtain most of their body heat from their surroundings. Ectotherms include fish, amphibians and reptiles.

Endotherms have a homeostatic mechanism to regulate their body temperature and obtain most of their body heat from their metabolism. The graph in Figure 3.22 shows the effect of external temperature on the body temperature of an endotherm and an ectotherm.

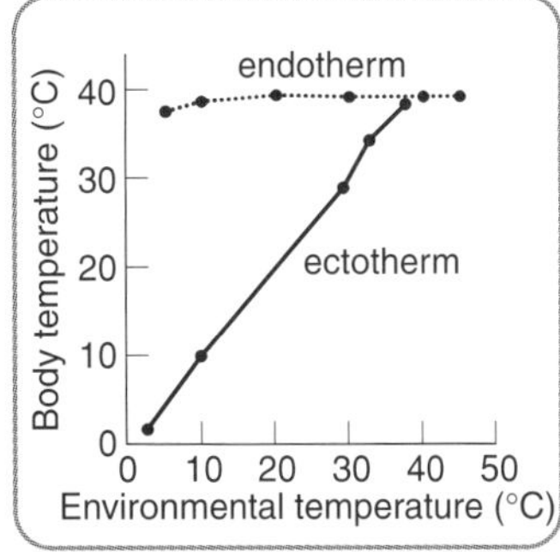

Figure 3.22 Effect of external temperature on body temperature

Enzymes control all the chemical reactions that take place in cells. They have an optimum temperature at which they work best. Animals, which can maintain the optimum temperature for enzyme activity, have a high metabolic rate and can function more efficiently.

The hypothalamus is the temperature-monitoring centre and contains thermoreceptors which detect changes in blood temperature. The hypothalamus sends out nerve impulses to the effectors, skin and body muscles.

Hot conditions, exercise or illness can increase the body temperature above normal. The thermoreceptors in the hypothalamus detect an increase in the body temperature and send nerve impulses to the effectors. The sweat glands increase sweat production. Heat from the body evaporates water in the sweat and lowers the body temperature. Vasodilation occurs and the skin arterioles become dilated (wider) allowing a larger volume of blood to flow through the capillaries on the skin surface. The blood loses heat by radiation.
These corrective mechanisms lower the body temperature to normal.

If the body temperature falls below normal the thermoreceptors in the hypothalamus detect the decrease and send nerve impulses to the effectors. The sweat glands decrease sweat production. Vasoconstriction occurs and the skin arterioles become constricted (narrower) reducing the volume of blood that flows through the capillaries on the skin surface. As a result, less heat is lost by radiation. The hair erector muscles contract to raise hairs. The trapped air provides insulation and reduces heat loss. Shivering of the skeletal muscles generates heat. There is an increase in the metabolic rate that increases heat production. These corrective mechanisms raise the body temperature to normal.

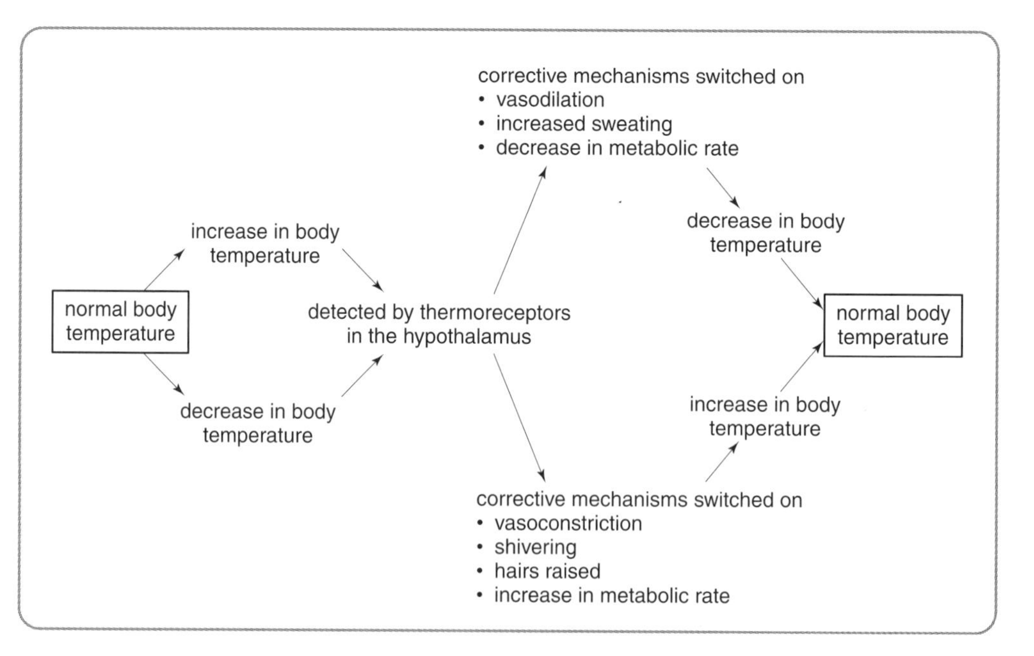

Figure 3.23 Summary of the control of body temperature in mammals

Top Tip

Try this for remembering the role of the hypothalamus as a monitoring centre.

H
Y
P
Osmoreceptors
Thermoreceptors
H
A
L
A
M
U
S

Exam Questions

Marks

SBQ 1 The flowchart below represents the homeostatic control of body temperature of a mammal.

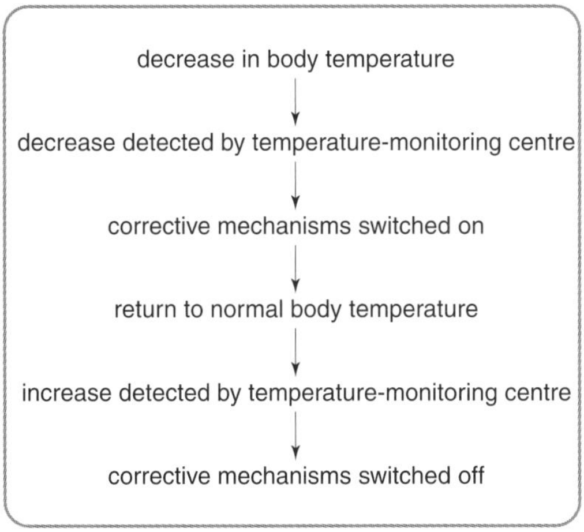

decrease in body temperature
↓
decrease detected by temperature-monitoring centre
↓
corrective mechanisms switched on
↓
return to normal body temperature
↓
increase detected by temperature-monitoring centre
↓
corrective mechanisms switched off

(a) Give the name of the temperature-monitoring centre in the brain that detects temperature changes. 1C

(b) In the following sentence, underline one of the alternatives in each pair to make the sentence correct.

***Exam Questions** continued* ➢

Exam Questions *continued*

In the **corrective mechanisms** given in the flowchart, (vasoconstriction/vasodilation) results in (increased/decreased) blood flow in the skin. 1C

(c) How are the instructions from the temperature-monitoring centre relayed to the corrective mechanisms in the skin? 1C

(d) State the importance of body temperature in humans to metabolic processes? 1B

SBQ 2 The flowchart below represents part of the homeostatic control of blood glucose concentration in a human.

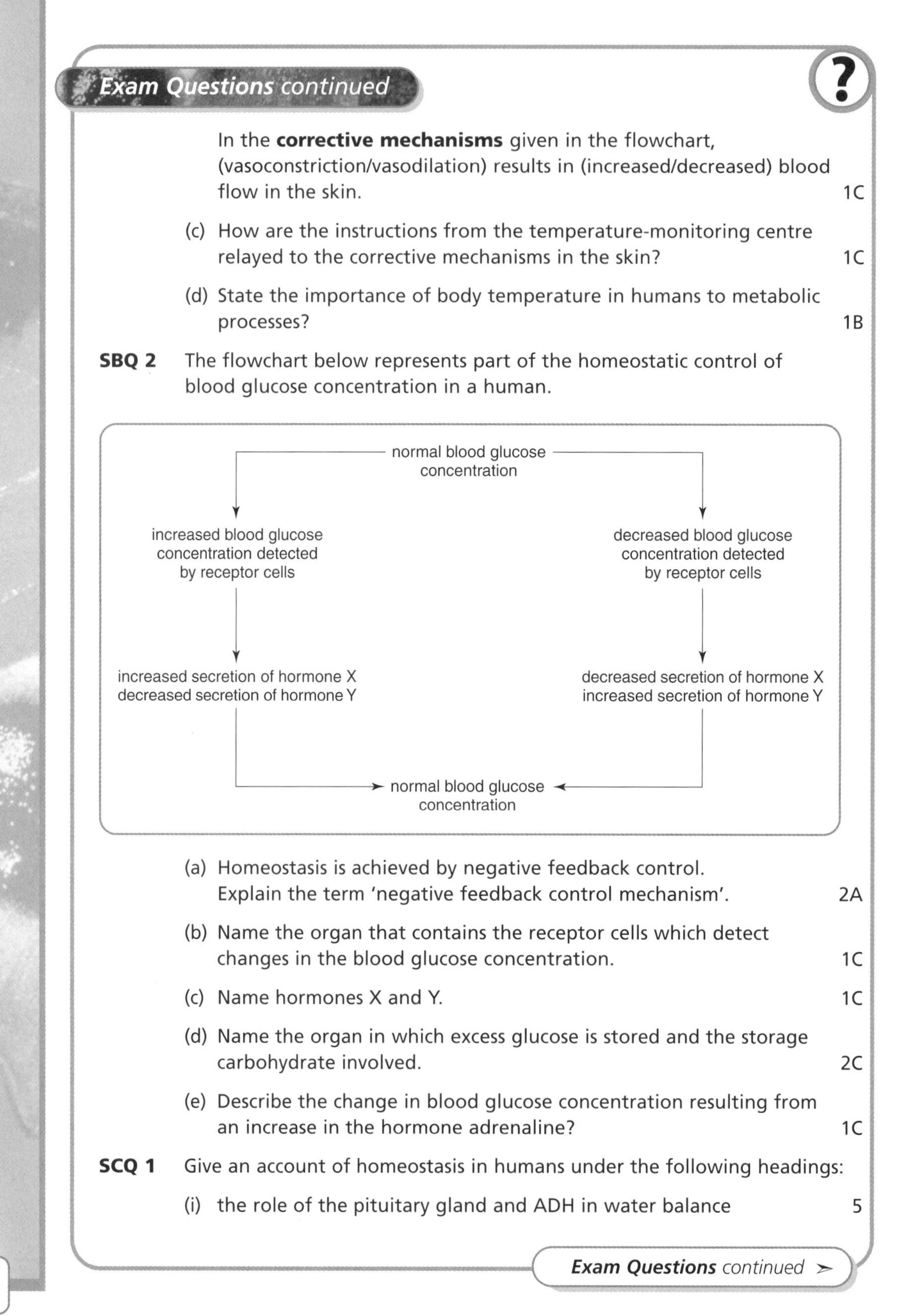

(a) Homeostasis is achieved by negative feedback control. Explain the term 'negative feedback control mechanism'. 2A

(b) Name the organ that contains the receptor cells which detect changes in the blood glucose concentration. 1C

(c) Name hormones X and Y. 1C

(d) Name the organ in which excess glucose is stored and the storage carbohydrate involved. 2C

(e) Describe the change in blood glucose concentration resulting from an increase in the hormone adrenaline? 1C

SCQ 1 Give an account of homeostasis in humans under the following headings:

(i) the role of the pituitary gland and ADH in water balance 5

Exam Questions *continued* ➢

(ii) the control of blood glucose concentration. 5

(10)

In question SCQ 2, ONE mark is available for coherence and ONE mark is available for relevance.

SCQ 2 Give an account of the mechanisms and importance of temperature regulation in endotherms. (10)

Answers on page 145–7

3.3 Population Dynamics

Key Ideas

☐ 1 A population is the number of individuals of the same species living in a habitat.

☐ 2 The number of individuals present per unit area or volume of a habitat is called the population density.

☐ 3 Population dynamics is the study of population changes and the factors that cause these changes.

☐ 4 Populations increase in number until they reach a size that the available resources in the environment can support. Despite short-term oscillations, the population then remains relatively stable.

☐ 5 The factors affecting the size of a population fall into two categories called density independent and density dependent factors.

☐ 6 Density independent factors are factors that affect the population regardless of the population density. Examples of density independent factors include extremes in temperature and rainfall and also natural disasters such as fire and volcanic activity.

☐ 7 The effect of density dependent factors increase as the population density increases.

☐ 8 Examples of density dependent factors include disease, food supply, predation and competition for space.

☐ 9 When food is in short supply competition occurs between individuals and some may starve to death.

☐ 10 If the population increases the predation also increases. Predators are more likely to kill a higher proportion of a dense population than a widely-spread one.

☐ 11 If the population increases then there is an increase in the spread of disease.

☐ 12 An increase in population results in increased competition for space or territory.

☐ 13 This competition will cause the population to decrease and return to the size that the environment can support.

☐ 14 Populations of plants and animals may be monitored to provide essential data that can be used for a wide variety of purposes. These include the management of food

***Key Ideas** continued ➢*

Key Ideas *continued*

and raw materials, the control of pest species, the assessment of pollution levels and the protection and conservation of endangered species.

- ☐ 15 Wild populations of human food species such as herring stocks or raw material species such as mahogany trees for timber must not be removed faster than they are able to reproduce or be replaced.
- ☐ 16 Population monitoring data allows sustainable management of these species such as the setting of quotas for fishing or harvesting.
- ☐ 17 Monitoring populations of pest species such as mosquitoes, rats, locusts and plant diseases provides information that allows effective control measures to be implemented.
- ☐ 18 Biological indicator species are organisms that, by their presence or absence, can be used to indicate levels of pollution. Lichens can be used as indicators of sulphur dioxide pollution and freshwater invertebrates such as mayflies indicate the level of dissolved oxygen in water.
- ☐ 19 Monitoring populations provides information on species that are endangered and in need of protection and conservation. It also provides data to measure the effectiveness of any control measures adopted.
- ☐ 20 Plant succession is the gradual change in the species of plants present in a particular habitat.
- ☐ 21 Succession is unidirectional and causes habitat modification.
- ☐ 22 The first species to move into an area are pioneer species such as lichens, mosses, herbs and grasses.
- ☐ 23 Pioneers change the habitat to allow further species to colonise.
- ☐ 24 Successive communities colonise one by one until the climax community is reached.
- ☐ 25 A climax community has a greater biomass, species diversity and stability and more complex food webs than previous communities.

Topic Notes

A population is the number of individuals of the same species that occupy a habitat. The number of individuals present per unit area or volume of a habitat is called the population density.

Populations increase in number until they reach a size that the available resources in the environment can support. The population then remains relatively stable despite short-term oscillations (rises or falls) of numbers. Changes in populations over time is called population dynamics.

The factors affecting the size of a population fall into two categories called density independent and density dependent factors.

Top Tip

Remember the **DIDD**y factors – **D**ensity **I**ndependent and **D**ensity **D**ependent factors.

Density independent factors affect the population regardless of the population density. Examples of density independent factors include extremes in temperature and rainfall that can bring about drought or floods.

The effect of density dependent factors increase as the population density increases. Examples of density dependent factors include disease, food supply, predation and competition for resources.

Top Tip

You should be able to relate all these factors to their **effect** on the population.

When food is in short supply, competition occurs between individuals and some may starve to death. If the population increases the predation also increases. Predators are likely to kill more of a dense population than a widely-spread one. If the population increases then there is an increase in the spread of disease. An increase in population results in increased competition for space or territory. This competition will cause the population to decrease and return to the size that the environment can support.

Populations of plants and animals may be monitored to provide essential data that can be used for a wide variety of purposes. These include the management of species used for food or raw materials.

Populations of species that provide humans with food such as herring and cod or raw materials such as trees for timber must not be reduced faster than they are able to reproduce or be replaced. Population data allows sustainable management of these species. For example, fish stocks can be controlled using fishing quotas and rainforest can be conserved using legislation.

Monitoring populations of pest species such as mosquitoes, rats, locusts and plant diseases provides information that allows effective control measures to be implemented.

Biological indicator species are organisms that, by their presence or absence, can be used to indicate the quality of the air or water in the environment. Species of lichens can be used as indicators of sulphur dioxide pollution and certain freshwater invertebrate species as indicators of the levels of dissolved oxygen in water.

Monitoring populations provides information on species such as the Siberian tiger that are endangered and in need of protection and conservation. It also provides data to measure the effectiveness of any control measures adopted.

Plant succession is the gradual but steady change in the diversity of species of plants present in a particular habitat. Succession is unidirectional and causes habitat modification.

The first species to move into an area are called pioneer species. These early colonisers are usually wind dispersed, have a low nutrient requirement, a shallow root system and are adapted to adverse conditions. Their presence increases the soil fertility and depth and improves its water retention, pH, drainage and aeration. After a short period of time a community makes the habitat less favourable for itself and more favourable for a different community that will therefore succeed it. This process is repeated until the climax community is reached.

As succession progresses the communities have greater biomass, species diversity and more complex food webs.

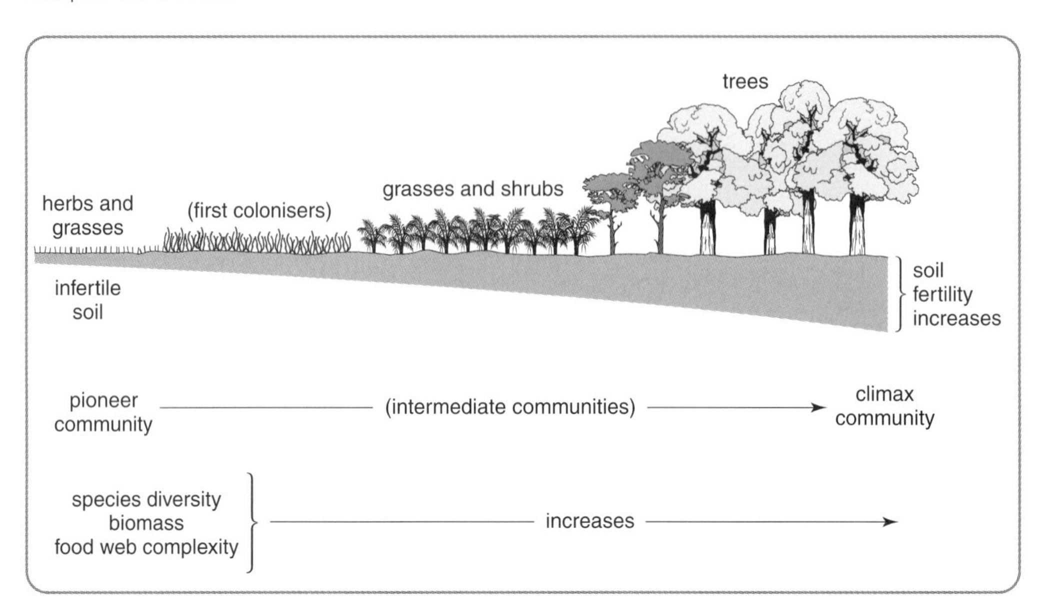

Figure 3.24 The main features of plant succession

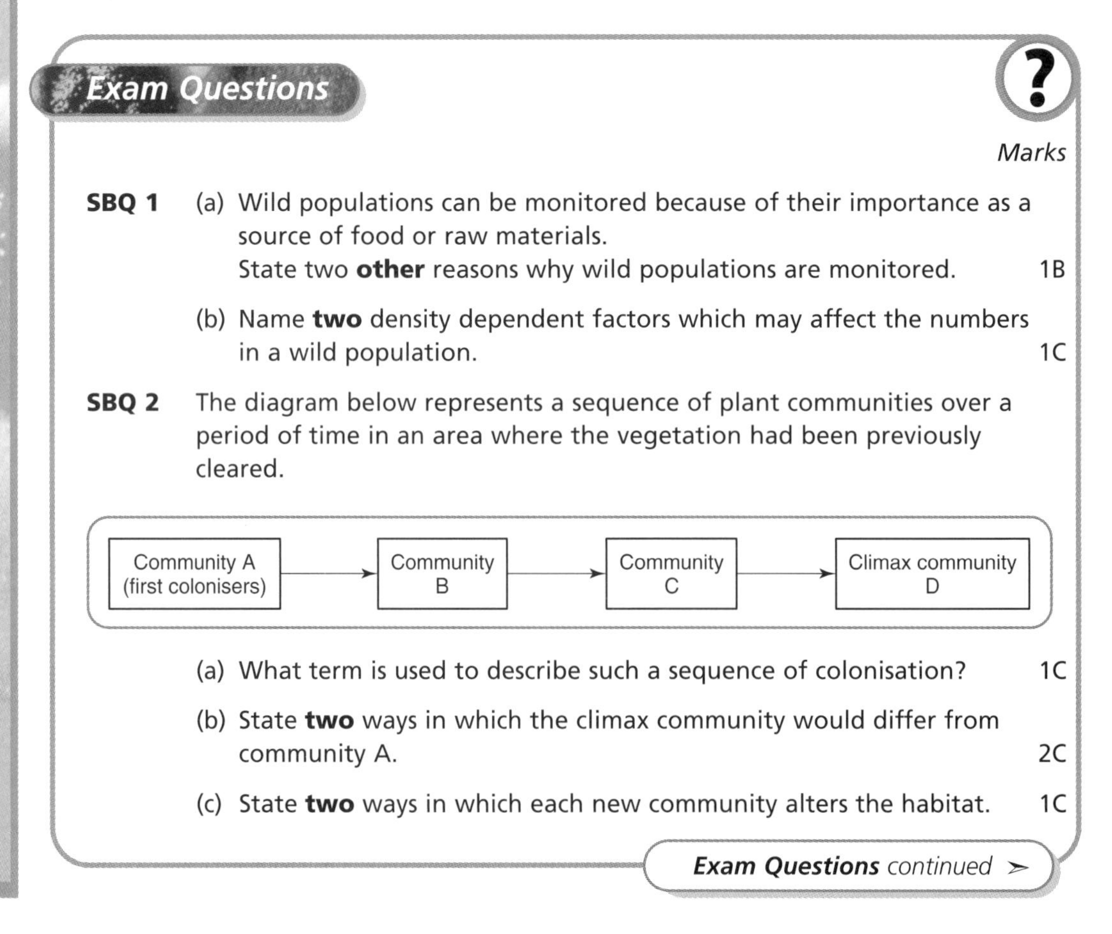

Exam Questions

Marks

SBQ 1 (a) Wild populations can be monitored because of their importance as a source of food or raw materials.
State two **other** reasons why wild populations are monitored. 1B

(b) Name **two** density dependent factors which may affect the numbers in a wild population. 1C

SBQ 2 The diagram below represents a sequence of plant communities over a period of time in an area where the vegetation had been previously cleared.

(a) What term is used to describe such a sequence of colonisation? 1C

(b) State **two** ways in which the climax community would differ from community A. 2C

(c) State **two** ways in which each new community alters the habitat. 1C

Exam Questions *continued* ➢

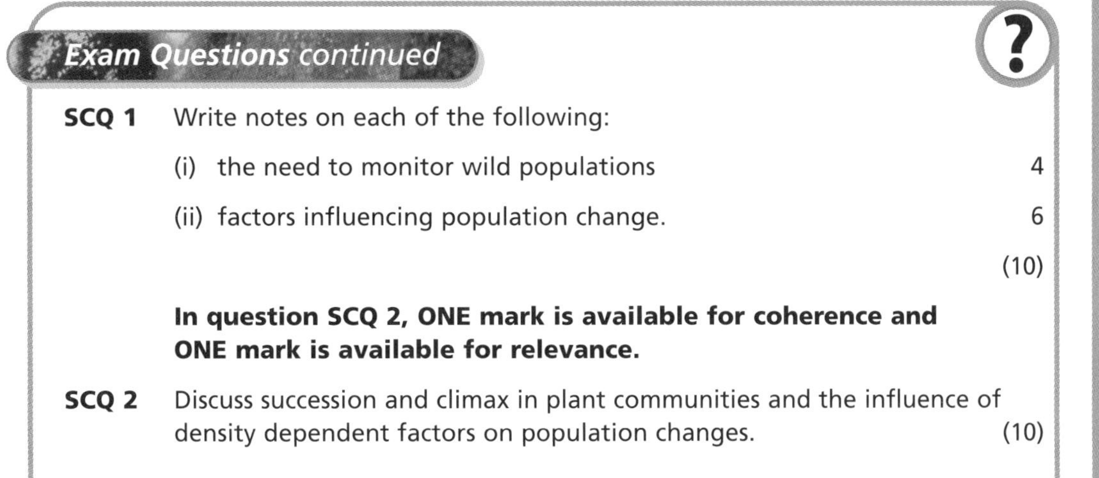

Exam Questions continued

SCQ 1 Write notes on each of the following:

(i) the need to monitor wild populations 4

(ii) factors influencing population change. 6

(10)

In question SCQ 2, ONE mark is available for coherence and ONE mark is available for relevance.

SCQ 2 Discuss succession and climax in plant communities and the influence of density dependent factors on population changes. (10)

Answers on page 148–50

Chapter 4

PROBLEM SOLVING – NOT A PROBLEM!

Many students ask,'What's the best way to revise for Higher Biology?' They often say, 'I feel I have studied really well and still, when I do past papers, I can't seem to get a high mark! I think it's mainly down to the Problem Solving questions. I'm having quite a bit of difficulty with them but I can't find any information on how to tackle these questions.'

The problem solving and practical skills that you will need and which will **definitely** be tested in your Course exam are all detailed in the Arrangements for Higher Biology.

Questions relating to each of the following points **will** be included in the exam. You **will** be asked to show that you have mastered each of these skills. We have numbered them P1–P7 for reference.

P1 Select relevant information from text, tables, charts, graphs and diagrams.

P2 Present information appropriately in a variety of forms such as tables or graphs.

P3 Process information accurately. Do calculations such as percentages, percentage changes, ratios and averages.

P4 Plan and design experimental procedures. Identify variables, controls and measurements required.

P5 Evaluate experimental procedures. Comment on such things as the control of variables, the limitations of the set-up or equipment, appropriateness of controls, sources of error and suggested improvements.

P6 Draw valid conclusions and give explanations supported by evidence.

P7 Make predictions based on available evidence.

Our aim in this section is to help you with these skills and provide you with techniques to score well in each of these types of questions. A few extra marks could make all the difference!

The Exam Questions in this chapter cover the skills required and the Answers and Commentary chapter, as well as the Top Tips, should help you to transfer your learning and skills to other questions of this type.

The most useful way to further develop these skills and improve your overall grade is to work through as many past paper questions as you can which cover these skills.

Search them out in each paper, identifying the different skills and trying each type.

Know the skills that you are good at and those that require further attention. By taking a more focussed approach you can sharpen your skills and achieve better grades.

Types of Skill

P1 Selecting information

The most commonly asked questions in this area require the candidate to select information from a graph. These questions usually start with statements such as:

From the graph, calculate the difference between…

OR

Use values from the graph to describe the effect of…

Top Tips

- Always read the introduction and the stem of the question carefully.
- Make yourself familiar with the variable being altered on the *x*-axis and what is being measured on the *y*-axis.
- Take note of the units used on each axis and be ready to use them in your later answers.
- Watch out at this level for graphs with a double *y*-axis.
  The two *y*-axes often have different scales to increase the difficulty for you. You must take great care to read the question and then the graph carefully to ensure that you are reading the correct *y*-axis. Candidates are so used to reading a graph from the usual left-hand *y*-axis that they often make a mistake.
- Calculate the divisions on the scales by working out the value of each small square. Again, take care with graphs that have a double *y*-axis in which the examiners deliberately give two different scales to see if you have this skill.
- If you are asked to calculate an increase or decrease between points on the graph then you should use a ruler to ensure accuracy and actually draw lines on the graph. This could avoid your eye skipping over a square and making a mistake.
- When you are asked to **describe** a trend, it is essential that you **quote values** at the appropriate points and use the exact labels or words given on the axes in your answer. This will usually involve giving four or five values and the correct units in your description.

P2 Presenting information

The most commonly asked questions in this area require candidates to present information that has been provided in a **table** in the form of a **graph** – usually a line graph.

Top Tips

- First things first. Check the instructions to find out if it is a line graph or a bar graph that is required.
- The marks are given for labelling the axes correctly, plotting the points and joining the points usually with straight lines using a ruler.

Top Tips continued ➢

Top Tips continued

- The graph labels should be identical to the data table headings and units. Copy them exactly, leaving **nothing** out.
- You need to decide which variable/table heading is to be plotted on each axis. The data for the variable being changed is placed on the *x*-axis. The left-hand table heading is usually the variable required for the *x*-axis. The right-hand table heading provides the label and data for the *y*-axis. You will lose a mark if these are reversed.
- You must select a suitable scale. It must use 50% or more of the axes provided otherwise a mark will be deducted. The value of the divisions on your scale should allow you to plot all of the points accurately.
- On your scales, make sure that you include the scale values **above** the highest values given on the data table.
- The scale must rise in regular steps. At Higher level, examiners often test you on this by having an irregular sequence of values in the data table. Watch out for this deliberate change in the data that is designed to check your care in plotting.
- Be careful to include one or both zeros at the origin if appropriate. It is acceptable for a scale to start with a value other than zero as long as none of the points to be plotted lie between your first value and zero.
- Take great care to plot the points accurately using an X or a dot (·) and then connect them exactly using a ruler.
- Do not plot 0 or connect the points back to the origin unless these values have actually been included in the data table. However, **if 0 is there you must plot it**.
- Things to consider when drawing a bar graph include ensuring that the bars are the same width and the tops of the bars are drawn with a ruler. You should be able to plot two groups of information on the same bar graph and remember to include a key.
- If you make a mistake in a graph, which cannot be neatly corrected, use the second grid provided at the end of your exam paper. Drawing your first graph in pencil could save time if a mistake is made.

P3 Processing information

The most commonly asked questions in this area are calculations such as percentages, ratios and averages.

Percentages

(a) Expressing a number as a percentage

The number needed as a percentage (%) is divided by the total and then multiplied by 100 as shown.

$$\frac{\text{Number needed as a \%}}{\text{Total}} \times 100$$

(b) Percentage change – increase or decrease

Do this calculation in two steps.

Step 1: calculate the change – increase or decrease.

Step 2: express this value as a %.

$$\frac{\text{Change (increase or decrease)}}{\text{Original starting value}} \times 100$$

Ratios

These questions usually require you to express the values given or being compared as a simple whole number ratio.

- First you need to obtain the values for the ratio from the data provided in the table or graph.
- Take care that you present the ratio values in the order they are stated in the question.
- Simplify them by dividing the larger number by the smaller one then dividing the smaller one by itself. If this does not give a whole number then you need to find another number that will divide into both of them.

For example 21 : 14 cannot be simplified by dividing 21 by 14 since this would not give a whole number. You must look for another number to divide into both, in this case 7. This would simplify the ratio to 3 : 2 that cannot be simplified any further.

Averages

These are similar to Standard Grade or Intermediate 2 calculations. Add up the values provided and then divide the total by the number of individual values given.

P4–P7 Experimental procedures

Experimental questions are specifically set to test your skills in this area and require you to:

- discuss aspects such as reliability, variables and fairness, validity, controls, measurements required, sources of error, suggested improvements
- draw conclusions
- make predictions.

The parts of the experimental question commonly take these forms:

- Give a reason why the experiment was repeated.
- Give one precaution taken to ensure that the results would be valid.
- Describe a suitable control.
- Explain why a control experiment was necessary.
- Why is it good experimental procedure to…?
- What precautions should be taken to minimise errors…?
- Why was the solution/experiment left for 20 minutes before…?

Top Tips

These tips all apply to the experimental setting question.

- To ensure the **reliability** of experiments and the results obtained, the experiment should be **repeated** and/or **many readings or samples** should be taken. Averages do not themselves increase reliability but are calculated to simplify the large quantities of repeat data.

  Remember **ROAR**! **R**epeat, **O**btain an **A**verage to increase **R**eliability.
- To ensure **validity**, only the variable being investigated should be altered while the other variables should be strictly controlled and kept constant.
- Examples of the variables which need to be controlled and kept constant to ensure results would be valid include temperature, pH, concentrations, mass, volumes, length, number, surface area and type of tissue. On no account use 'amount' – do not write about the **amount** of anything!
  Usually the question asks for variables 'not already mentioned ' meaning that you are looking for variables not already given in the question stem.
- A control experiment allows a comparison to be made and allows you to attribute any change or difference in the results to the factor or variable being altered.
- The control should be identical to the original experiment apart from the one factor being investigated.
  If you are asked to describe a suitable control make sure that you **describe it in full**. It is not enough to say 'keep everything the same except for the one variable being changed'. Give a full description of the control including quantities if appropriate.
- Questions regarding procedure that ask why an experiment was left for a certain time require you to state that this is to allow enough time for a specific process or event to occur. For example, enough time for:

  ... diffusion or absorption of chemicals into tissue.
  ... growth to take place.
  ...the effect of applied chemicals to be seen.
  ...the reaction to proceed.

  These are general comments but try to be specific where possible and use the named chemicals or reactions that relate to the particular question.
- If the effect of temperature on enzyme activity is being investigated, it is good practice to allow the solutions to reach the required temperature before mixing them. If this is not done the reactions would be taking place at a different temperature and the results would be invalid.
- It is often good experimental practice to express results as a percentage change. A percentage change allows a fair comparison to be made when, for example, original lengths or masses are different.
- Watch out for the questions that refer to dry mass of tissues.
  Since the water content of tissues is variable and can change from day to day, the dry mass is often used when referring to growth or when comparing tissues. This allows a fair comparison to be made or, in the case of growth studies, increases the validity of the results.

Top Tips *continued* ➢

Top Tips continued

- Precautions to minimise errors include washing apparatus such as beakers or syringes or using different ones if the experiment involves different chemicals or different concentrations. This prevents **cross** contamination.
- Don't worry if the practical setting is unfamiliar; the general principles are always the same.
- In drawing conclusion from results it is usually necessary to spot a pattern or trend in the results and relate this to the aim of the investigation. If the investigation is into enzyme activity the results pattern would need to be related to changes in the activity of the enzyme.
- When predictions are asked for, it is often necessary to extend results in a graph or table to values of the variable that have not actually been tested. It is acceptable to extend graph lines on an answer graph to allow this to be done.

Exam Questions

1 Example of a Data Handling question with a high level of demand *Marks*

(i) Graph 1 below shows how the pulse rate and stroke volume, which is the volume of blood pumped out of the heart in one beat, change with the level of exercise.
The level of exercise is measured as rate of oxygen uptake.

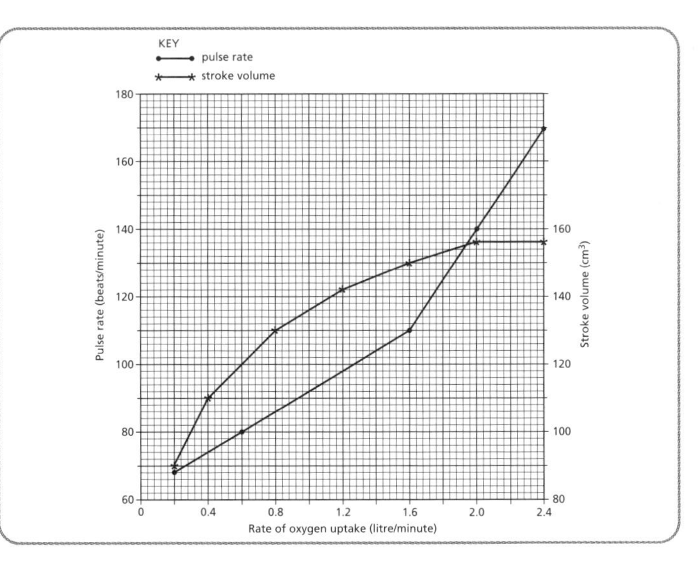

Graph 1

(a) What is the pulse rate and stroke volume when the rate of oxygen uptake is 1.4 litres/minute? 1B

Exam Questions continued ➢

Exam Questions continued

(b) Use values from Graph 1 to describe the changes in pulse rate and stroke volume when the rate of oxygen uptake increased from 1.6 to 2.4 litres/minute. **2B**

(c) Calculate (in litres) the total volume of blood leaving the heart in one minute when the rate of oxygen uptake is 0.4 litres/minute. **1**

(d) Calculate the increase in the rate of oxygen uptake when the pulse rate is increased from 80 beats/minute to 130 beats/minute. **1**

(e) Calculate the percentage increase in the rate of oxygen uptake when the pulse rate increases from 110 to 140. **1**

(f) Express as the simplest whole number ratio, the pulse rate to stroke volume when the oxygen uptake is 0.6 litres/minute. **1**

(ii) Ventilation rate is calculated as the volume of air inhaled during one minute.

Graph 2 shows how the ventilation rate changes with the level of exercise.

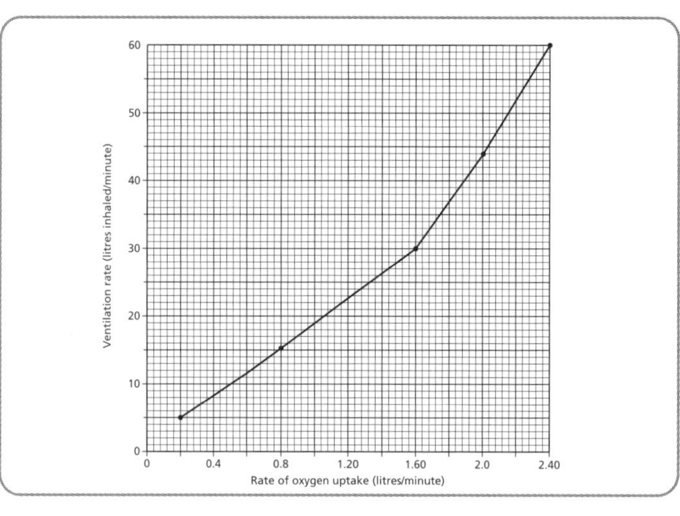

Graph 2

(a) Given that 20% of the air is oxygen, calculate the volume of oxygen inhaled per minute when the rate of oxygen uptake is 1.6 litres/minute. **1B**

(b) Using information from Graph 1 and Graph 2, calculate the ventilation rate when the pulse rate is 110 beats/minute. **1A**

Exam Questions *continued* ➢

Exam Questions continued

2 Example of an Experimental Procedure question

An investigation was carried out into the effects of osmosis on potato tissue.
Pieces of potato were placed in different concentrations of salt solutions for one hour.
The results are shown in the table below.

Concentration of salt solution (M)	Mass of potato at start (g)	Mass of potato after 1 hour (g)	Percentage change in mass (%)
0.05	4.0	4.8	+ 20
0.10	3.5	4.2	+ 20
0.20	4.4	4.7	+ 7
0.25	3.7	3.7	0
0.35	3.9	3.4	– 13

(a) Plot a line graph on a piece of graph paper to show the percentage change in mass of the potato pieces against concentration of salt solution. 2C/B

(b) Identify two variables that must be controlled to make the results obtained valid. 1C

(c) The pieces of potato were left in the salt solution for one hour. Explain why this was good experimental procedure. 1B

(d) Each piece of tissue was gently blotted in a paper towel before each time it was weighed.
Suggest why this was done. 1B

(e) From the information given, why was it good experimental procedure to use percentage change in mass when comparing results? 1B

(f) Give one way in which the reliability of the results could have been improved. 1C

(g) Draw a conclusion from the results in the table. 1C

(h) Predict the percentage change in mass of piece of potato placed in a 0.40 M salt solution for one hour. 1C

Answers on page 151–3

Chapter 5

WRITING EXTENDED RESPONSES – A KEY TO BETTER GRADES!

General

This is quite simply the toughest part of the paper – but it's easy to improve your performance. Remember though, **there is no substitute for knowledge.**

The secret of good extended writing is practice. **There is also no substitute for practice**. It improves your knowledge, your technique and your timing.

We suggest that you try each extended response in the Exam Questions sections of this book, mark these yourself using the SQA standard answers in the Answers and Commentary chapter and make corrections to give yourself a set of excellent fully correct responses.

We believe that it is worth having a few extended responses off pat. Pick out a few – say two from each unit – which you basically memorise. If one appears, even in part, then you're on a real winner. If none appears, the knowledge will come in useful somewhere else in the paper – you cannot lose!

Choosing the Best Question to Answer

Each extended response has a choice of two possible questions. Getting the right choice is important. Examiners find that in each paper one of the options in each section tends to be better answered across the country than the other. Give yourself a good few minutes to make your choice for each question.

It is not a good idea to change your mind after you have written a response, but if you have to, don't score out your original answer until you are certain your second try is better.

Diagrams

Diagrams are clearly useful as they can save time and help to join ideas together. They are totally useless without labels and annotation. **The marks are in the labels.**

If your drawings are in a sequence like a flow chart or in a time order, remember to put arrows to show direction – marks are easily lost if you don't.

Some answers almost demand diagrams – its tough to describe a chloroplast or a plasma membrane without drawing it!

Question 1

This question is structured into parts to guide your answer selection. The marks available for each part are given and these provide a clue to the length of answer required.

Question 2 Coherence and Relevance

These two marks must not be missed! If you can't answer the question you will not get these marks in any case but if you can even pick up four or five marks out of the eight for content it is crazy not to gain the coherence and relevance marks too.

- **Coherence** To get this mark you should **use sub-headings from the question** – it's as simple as that.
- **Relevance** To get this mark **do not mention irrelevant points**. If the question is about endotherms do not even mention ectotherms; if the question is about DNA do not even mention RNA; if the question is about aerobic processes do not even mention anaerobic processes – enough said?

Chapter 6

EXAMINATION TIPS – A WAY TO EXTRA MARKS!

Section A

Summary

30 Multiple Choice items – 30 marks

- Do not spend more than **30 minutes** on this section.
- Answer on a grid. Make sure the grid has your name pre-printed on it.
- **Do not leave blanks** – complete the grid for each question as you work through.
- Try to answer each question in your head **without** looking at the options. If your answer is there – you are home and dry!
- If not certain, choose the answer that seemed most attractive on **first** reading the answer options.
- If you are guessing, try to eliminate options before making your guess. If you can eliminate three – you are left with the correct answer even if you do not recognise it!

Section B

Summary

80 marks of Structured Questions – 80 marks

- Spend about **85 minutes** on this section.
- Answer on the question paper. Try to write neatly and keep your answers on the support lines if possible – the lines are designed to take the full answer! Another clue to answer length is the mark allocation – if there are 2 marks available your answer will need to be longer and may well have two, three or even four parts.
- The questions are usually laid out in Unit sequence but remember some questions are **designed** to cover more than one Unit.
- The C-type questions usually start with '**State**', '**Give**' or '**Name**'.
- Questions that begin with '**Explain**' and '**Describe**' are usually A-type and are likely to have more than one part to the full answer.
- Abbreviations like DNA and ATP are fine but remember you must name the nucleotide bases in full – A, T, G and C won't get the marks.

***Summary** continued* ➢

Summary continued

- Don't worry that the data and practical questions are in unfamiliar contexts, that's the idea! Just keep calm and read the questions carefully.
- Remember to '**use values from the graph**' if you are asked to do so.
- Draw graphs using a ruler and use the data table headings for the axes labels.
- Look out for graphs with 2 *y*-axes – these need extra special concentration and anyone can make a mistake!
- Answers to calculations will not usually have more than three decimal places.
- If there is a space for calculation given – you will very likely need to use it!
- Do not leave blanks. Have a go using the language in the question if you can.

Section C

Summary

Two Extended Response questions – 20 marks

- Spend about **35 minutes** on this section.
- Best to do this section last after you have been thinking Biology for nearly two hours.
- Give yourself a few minutes to make your question choice – this is very important.
- Use the titles and sub-headings given on the question paper. In the second extended response question, sub-headings or paragraphs are **essential**.
- Diagrams are often useful but they **must** be fully labelled.

Chapter 7

EXAM QUESTIONS: ANSWERS AND COMMENTARY – THE SQA STANDARD!

1.1 Cell Structure in Relation to Function

Answers

Answer to SBQ 1

V, D, T, C, A, B

3/2 correct	= 1	
5/4 correct	= 2	
All correct		3

Answer to SBQ 2

Can detect light/has a light detector

Has chloroplasts

Has a flagellum to move to light

2 ideas correct	= 1	
All three ideas correct		2

Three ideas for 2 marks makes this an A-type question.

Answer to SBQ 3

W, P, P, P, W

All correct	2
4/3 correct	1

Answer to SBQ 4

(a)	X protein	
	Y phospholipid	
	Both correct	1
(b)	A active transport	
	B diffusiion/osmosis	
	Both correct	1

Answers continued

Answer to SBQ 5

Oxygen is needed for respiration/energy release/ATP production	= 1	
Energy/ATP needed for active uptake	= 1	2

Top Tip

See how the CORE facts can help gain marks?

An A-type question requiring a ***link*** *between respiration and active transport*

Answer to SCQ 1

(i)	1	Double phospholipid layer	1
	2	Protein embedded/proteins form a mosaic.	1
	3	Proteins form pores.	1
	4	Membrane is fluid/phospholipids are in constant motion.	1
		Any to a maximum of 3 marks	**3**
(ii)	5	Made of cellulose fibres	1
	6	Gives support to cell	1
	7	Fully permeable	1
	8	Prevents bursting of cell when turgid	1
		Any to a maximum of 3 marks	**3**
(iii)	9	Movement from low to high concentration	1
	10	Energy/ATP needed	1
	11	Aerobic respiration provides energy/ATP	1
	12	Protein carriers in plasma membrane involved	1
	13	Allows selective ion uptake	1
		Any to a maximum of 4 marks	**4**
			10

Answer to SCQ 2

1	Double phospholipid layer	1
2	Protein embedded/protein forms a mosaic.	1
3	Protein forms pores.	1
4	Membrane is fluid/phopholipids are in constant motion.	1
	Any to a maximum of 3 marks	**3**

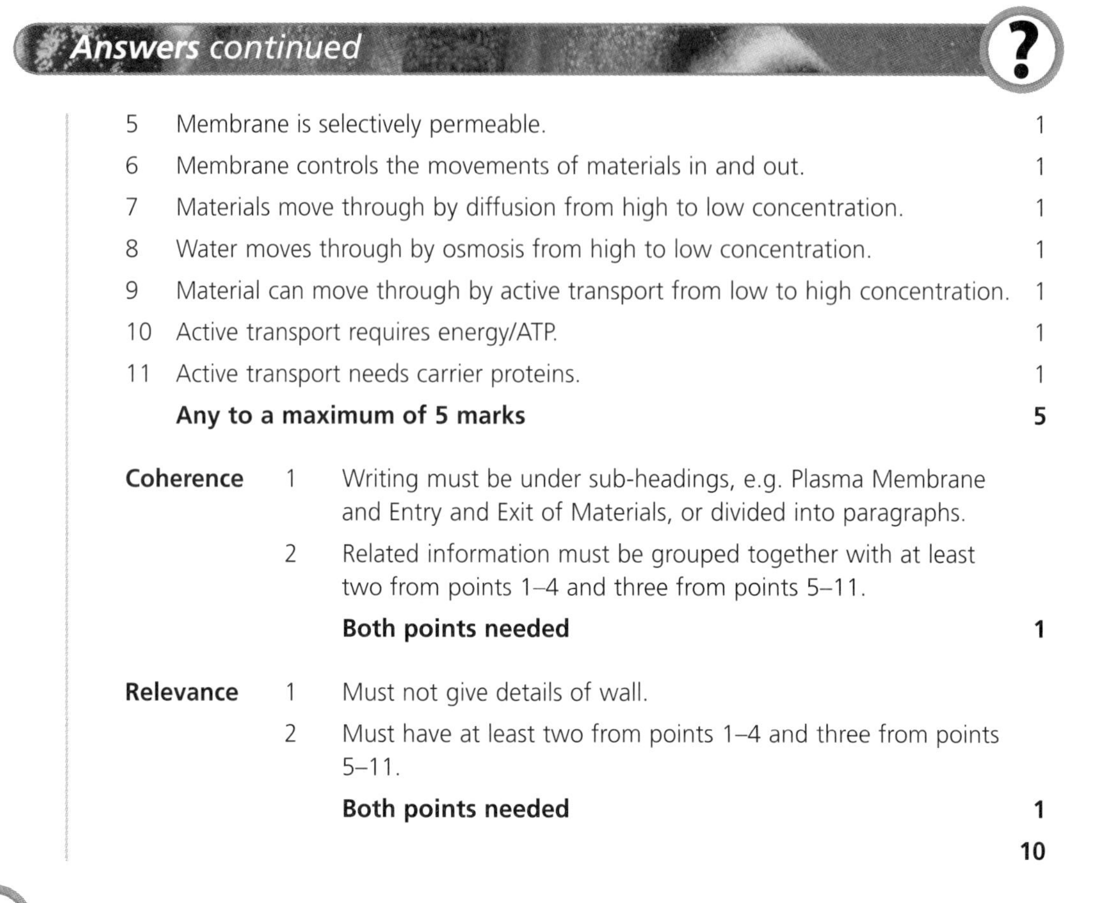

Answers continued

5	Membrane is selectively permeable.	1
6	Membrane controls the movements of materials in and out.	1
7	Materials move through by diffusion from high to low concentration.	1
8	Water moves through by osmosis from high to low concentration.	1
9	Material can move through by active transport from low to high concentration.	1
10	Active transport requires energy/ATP.	1
11	Active transport needs carrier proteins.	1
	Any to a maximum of 5 marks	**5**

Coherence	1	Writing must be under sub-headings, e.g. Plasma Membrane and Entry and Exit of Materials, or divided into paragraphs.	
	2	Related information must be grouped together with at least two from points 1–4 and three from points 5–11.	
		Both points needed	**1**
Relevance	1	Must not give details of wall.	
	2	Must have at least two from points 1–4 and three from points 5–11.	
		Both points needed	**1**
			10

1.2 Photosynthesis

Answer to SBQ 1

(a)	Absorbed, reflected or transmitted	
	Any two	2

Top Tip

Remember **ART**?

(b)	Grana/granum	1
(c)	NADPH and ATP	
	Both	1
(d)	Stroma	1
(e)	1, 3, 5, 6	
	Two or three correct = 1	
	All correct	2

Answers continued

Answer to SCQ 1A

(i)	1	Grana are stacks of membranes.	1
	2	Grana contain the photosynthetic pigments.	1
	3*	Light dependent stage of photosynthesis takes place in the grana.	1
	4	Stroma is the fluid filled region surrounding the grana.	1
	5	Stroma contains enzymes.	1
	6	Carbon fixation stage/Calvin cycle takes place in the stroma.	1
		Any to a maximum of 4 marks	**4**
(ii)	7*	Light dependent stage of photosynthesis takes place in the grana.	1
	8	Pigments/chlorophyll absorb or trap the light energy.	1
	9	Several pigments allow absorption over a wide spectrum or increase absorption over a wider range of wavelengths.	1
	10	Chlorophyll is the main pigment and mainly absorbs in the blue and red regions of the spectrum.	1
	11	Accessory pigments pass energy to the chlorophyll.	1
	12	Some of the energy is used to split water into hydrogen and oxygen/ some of the energy is used in the photolysis of water.	1
	13	Hydrogen released combines with the hydrogen acceptor NADP to form NADPH.	1
	14	Oxygen is released as a by-product.	1
	15	Some of the energy is used to synthesise/produce/regenerate ATP.	1
		Any to a maximum of 6 marks	**6**
*award only once			**10**

Answer to SCQ 1B

(i)	1	Grind or mash the leaves with acetone.	1
	2	Filter or centrifuge the extract to remove the cell debris.	1
	3	Repeat the spotting procedure or applications on the chromatography paper or thin layer.	1
	4	Allow the solvent time to run.	1
	5	Pigments travel different distances or are separated at different rates.	1
	6	Separation order is first carotene which is the most soluble and travels furthest followed by xanthophyll, chlorophyll a and then chlorophyll b.	1
		Any to a maximum of 4	**4**
(ii)	1	Carbon fixation stage occurs in the stroma.	1
	2	Controlled by enzymes or it is an enzyme controlled sequence of reactions.	1
	3	RuBP (5C) is the carbon dioxide acceptor.	1

Answers continued

4	Forms an unstable 6C compound which splits into two molecules of GP (3C).	1
5	GP (3C) is reduced or GP (3C) joins with hydrogen.	1
6	Hydrogen is supplied by the NADPH produced in the light dependent stage.	1
7	Energy to do this is supplied by the ATP produced in the light dependent stage.	1
8	GP (3C) is converted into glucose (6C).	1
9	Some GP is used to regenerate RuBP.	1
	Any to a maximum of 6 marks	**6**
		10

Answer to SCQ 2

1	Grana are stacks of membranes.	1
2	Grana contain the photosynthetic pigments.	1
3	Light dependent stage of photosynthesis takes place in the grana.	1
4	Stroma is the fluid filled region surrounding the grana.	1
5	Stroma contains enzymes.	1
6	Carbon fixation stage/Calvin cycle takes place in the stroma.	1
	Any to a maximum of 4 marks	**4**
7	Grind or mash the leaves with acetone.	1
8	Filter or centrifuge the extract to remove the cell debris.	1
9	Repeat the spotting procedure or applications on the chromatography paper or thin layer	1
10	Allow the solvent time to run.	1
11	Pigments travel different distances or are separated at different rates.	1
12	Separation order is first carotene which is the most soluble and travels furthest followed by xanthophyll, chlorophyll a and then chlorophyll b.	1
	Any to a maximum of 4 marks	**4**

Coherence	1	Writing must be under sub-headings e.g. Structure of Chloroplast and Extraction of Pigments or divided into paragraphs.	
	2	Related information must be grouped together with at least 2 points from each section.	
		Both points needed	**1**
Relevance	1	Must not give details of other organelles or light dependent stage.	
	2	Must have at least 2 points from each section.	
		Both points needed	**1**
			10

Answers continued

1.3 Energy Release

Answer to SBQ Q1

(a)	Fats/proteins	1
	This often turns out a A-type question as this could be a less emphasised area of the Arrangements.	
(b)	Cytoplasm	1
	Matrix of the mitochondria	1
(c)	2 **more** molecules gained than lost in the process.	1
(d)	Pyruvic acid 3C	1
	Acetyl group 2C	1
(e)	NAD	
(f)	Oxygen	1
	Water	1
(g)	Lactic acid	1

Answer to SCQ 1

(i)	1	Occurs in cytoplasm.	1
	2	Glucose broken down into 2 molecules of pyruvic acid.	1
	3	ATP required to start the process.	1
	4	Net gain of 2 ATP.	1
	5*	Hydrogen carried away by NAD.	1
	6	Does not requires oxygen.	1
		Any to a maximum of 3 marks	**3**
(ii)	7	Occurs in mitochondria matrix.	1
	8	Pyruvic acid converted to acetyl-CoA.	1
	9	4C compound joins acetyl group to form citric acid.	1
	10	Carbon dioxide produced as a waste product.	1
	11*	Hydrogen carried away by NAD.	1
	12	Requires oxygen.	1
		Any to a maximum of 4 marks	**4**
(iii)	13	Occurs on the mitochondria cristae.	1
	14	Hydrogen passed through a series of carriers.	1
	15	Oxygen is the final accep;tor of hydrogen.	1
	16	Water is the final metabolic product.	1
	17	Most ATP produced at this stage.	1
		Any to a maximum of 3	**3**

*award only once **10**

Answers continued

Answer to SCQ 2

1	Occurs in cytoplasm.	1
2	Glucose broken down into 2 molecules of pyruvic acid.	1
3	ATP required to start the process.	1
4	Net gain of 2 ATP.	1
5*	Hydrogen carried away by NAD.	1
6	Does not requires oxygen.	1
	Any to a maximum of 4 marks	**4**
7	Occurs in mitochondria matrix.	1
8	Pyruvic acid converted to acetyl-CoA.	1
9	4C compound joins acetyl group to form citric acid.	1
10	Carbon dioxide produced as a waste product.	1
11*	Hydrogen carried away by NAD.	1
12	Requires oxygen.	1
	Any to a maximum of 4 marks	**4**

*award only once

Coherence

1 Writing must be under sub-headings e.g. Glycolysis and The Krebs Cycle or divided into paragraphs.

2 Related information must be grouped together with at least two from points 1–6 and two from points 7–12.

Five marks must be scored in total.

Both points needed **1**

Relevance

1 Must not give details of cytochrome system or anaerobic respiration.

2 Must have at least two from points 1–6 and two from points 7–12 with a total of at least 5.

Both points needed **1**

10

1.4 Synthesis and Release of Proteins

Answer to SBQ 1

(a)	The DNA molecule unwinds/unzips.	1
(b)	Phosphate, deoxyribose (Both correct).	1
(c)	Hydrogen.	1
(d)	Cytosine; guanine; thymine Two correct = 1 mark All three correct	2

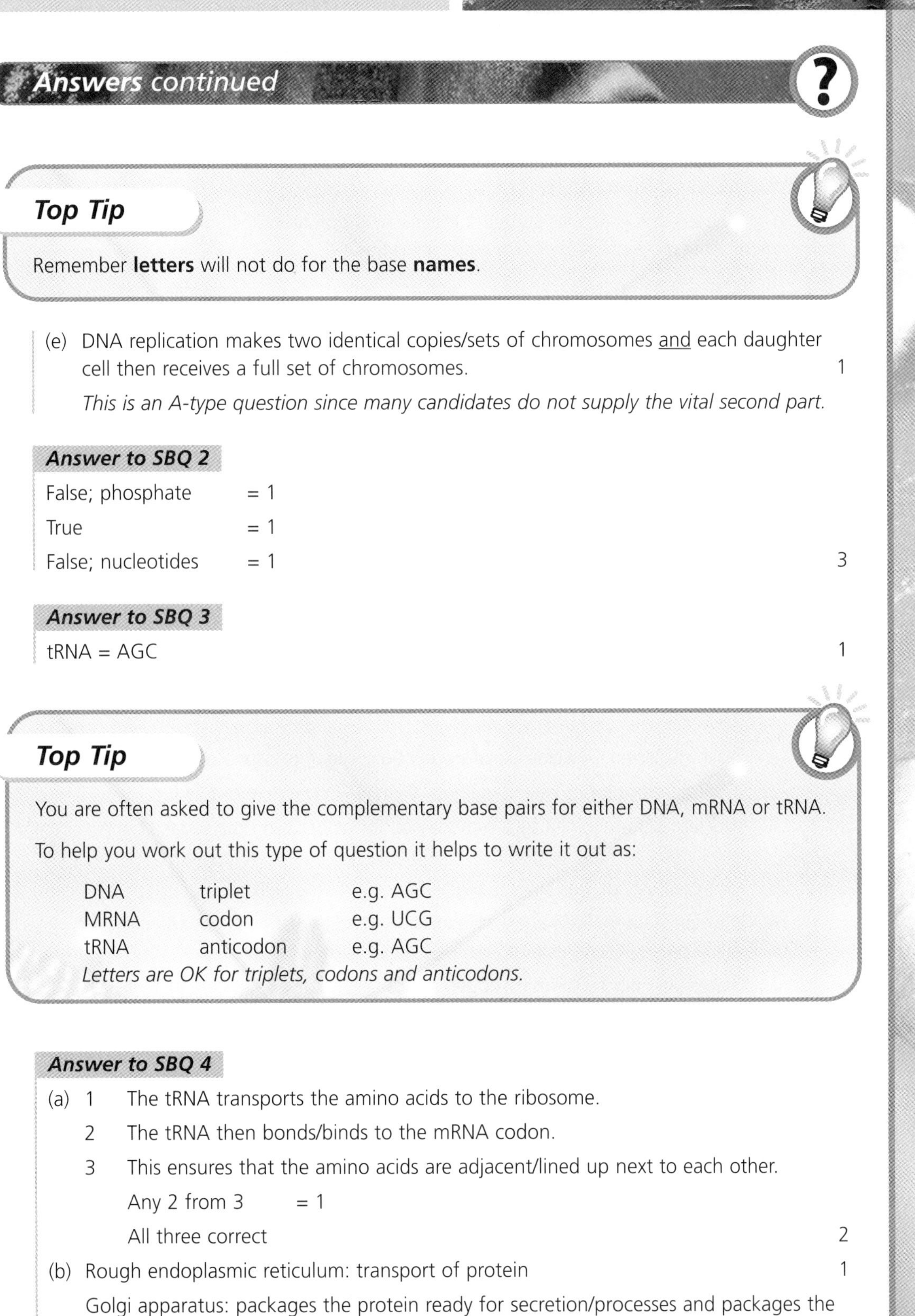

Answers continued

Top Tip

Remember **letters** will not do for the base **names**.

(e) DNA replication makes two identical copies/sets of chromosomes and each daughter cell then receives a full set of chromosomes. 1

This is an A-type question since many candidates do not supply the vital second part.

Answer to SBQ 2

False; phosphate = 1

True = 1

False; nucleotides = 1 3

Answer to SBQ 3

tRNA = AGC 1

Top Tip

You are often asked to give the complementary base pairs for either DNA, mRNA or tRNA.

To help you work out this type of question it helps to write it out as:

DNA	triplet	e.g. AGC
MRNA	codon	e.g. UCG
tRNA	anticodon	e.g. AGC

Letters are OK for triplets, codons and anticodons.

Answer to SBQ 4

(a) 1 The tRNA transports the amino acids to the ribosome.

2 The tRNA then bonds/binds to the mRNA codon.

3 This ensures that the amino acids are adjacent/lined up next to each other.

Any 2 from 3 = 1

All three correct 2

(b) Rough endoplasmic reticulum: transport of protein 1

Golgi apparatus: packages the protein ready for secretion/processes and packages the protein/adds carbohydrate to the protein 1

Answers continued

Answer to SBQ 5

(a) Fibrous (for collagen)

For globular, the examples stated in the learning outcomes include enzymes, membrane proteins, some hormones and antibodies.

Both 1

(b) 1 DNA is double stranded while RNA is single stranded.

2 DNA contains the sugar deoxyribose while RNA contains the sugar ribose.

3 DNA includes the base thymine but in RNA it is replaced with uracil.

Any 1 = 1

All three correct 2

Top Tip

A **full** comparison of DNA and RNA is always **best**.

Answer to SCQ 1

(i) 1 DNA is composed of nucleotides. 1

2 Each nucleotide is made up of deoxyribose sugar, phosphate and a base. 1

3 Four different bases called adenine, thymine, cytosine and guanine. 1

4 Nucleotides are joined by a bond which forms between the sugar of one nucleotide and the phosphate of another. 1

5 DNA consists of two strands twisted into a double helix. 1

6 Complementary base pairs which connect the two strands are adenine with thymine and guanine with cytosine. 1

7 Bases are held by weak hydrogen bonds. 1

Any to a maximum of 5 marks **5**

(ii) 8 DNA unwinds and unzips. 1

9 Enzymes and ATP are required. 1

10 Original DNA strands act as a template. 1

11 Free DNA nucleotides present form complementary base pairs i.e. adenine to thymine and cytosine to guanine. 1

12 The nucleotides join up forming a bond between the sugar of one nucleotide and phosphate of another. 1

13 Each of the DNA double strands contains one of the original strands. 1

14 Newly formed DNA molecules now retwist and form two double helixes. 1

Any to a maximum of 5 marks **5**

10

Answers continued

Answer to SCQ 2

1	DNA/gene unzips or the hydrogen bonds break.			1
2	Enzymes and ATP are required for the synthesis of mRNA.			1
3	RNA nucleotides form complementary base pairs with the bases of one strand of the DNA.			1
4	Bases pair as follows:	DNA	RNA	
		Adenine	Uracil	
		Thymine	Adenine	
		Guanine	Cytosine	
		Cytosine	Guanine	1
5	Nucleotides join up forming a bond between the sugar of one nucleotide and phosphate of another.			1
6	The mRNA is copied or transcribed from the DNA.			1
7	The mRNA passes out through the pores of the nucleus into the cytoplasm and attaches to a ribosome.			1
	Any to a maximum of 5 marks			**5**
8	tRNA molecules transport amino acids to the ribosome.			1
9	There is a specific tRNA for each amino acid.			1
10	Triplets of bases on mRNA form codons and the triplets of bases on tRNA form anticodons.			1
11	tRNA anticodons match to the mRNA codons.			1
12	Peptide bonds form between adjacent amino acids.			1
	Any to a maximum of 3 marks			**3**

Coherence	1	Writing must be under sub-headings e.g. mRNA Synthesis and Role of mRNA or divided into paragraphs.	
	2	Related information must be grouped together with at least two points from each section.	
		Both points needed	**1**
Relevance	1	Must not give details of DNA replication or protein secretion.	
	2	Must have at least two relevant points from each section.	
		Both points needed	**1**
			10

Answers continued

1.5 Cellular Response in Defence

Answer to SBQ 1

(a)	Stage 3: The viral nucleic acid/DNA/RNA is copied or replicated.	1
	Stage 6: Large numbers of viruses are released when the cell bursts open/lysis or cell bursts or ruptures or viruses are released from cell.	1
(b)	Amino acids, ATP, nucleotides or enzymes	
	Any 2	1
(c)	Ribosome	1

Answer to SBQ 2

(a)	Antibodies	1
(b)	Antigen or surface protein or foreign protein or protein coat	1

Answer to SBQ 3

(a)	Phagocytosis	1
(b)	Lysosomes	1

Answer to SBQ 4

(a)	Tannin, nicotine or cyanide	
	Any 2	1

Top Tip

Remember **T**hree **N**asty **C**hemicals?

(b)	Resin	1

Answer to SCQ 1

(i)	1	Phagocytosis is a cellular defence mechanism in animals.	1
	2	Membrane surrounds or engulfs bacteria.	1
	3	Bacteria are enclosed in a vacuole.	1
	4	Lysosomes attach to the vacuole.	1
	5	Enzymes digest the bacteria.	1
		Any to a maximum of 4 marks	**4**
(ii)	6	Antibodies are produced by lymphocytes.	1
	7	Antibodies are produced in response to foreign antigens.	1
	8	Antibodies are specific.	1

Answers continued

	9	Antibodies destroy or inactivate the antigens.	1
		Any to a maximum of 3 marks	**3**
(iii)	10	Plants can protect themselves by producing a variety of toxic compounds.	1
	11	Examples of toxic compounds include tannins, nicotine and cyanide. Any 2. One mark each.	2
	12	Some plants can isolate injured areas by means of substances such as resin.	1
		Any to a maximum of 3 marks	**3**
			10

Answer to SCQ 2

1	Viruses are very small.	1
2	They contain a nucleic acid / DNA or RNA which is surrounded by a protein coat.	1
3	They attack or infect specific host cells.	1
4	They can only replicate or reproduce inside host cells.	1
	Any to a maximum of 2 marks	**2**
5	Virus attaches to the host cell.	1
6	Virus/nucleic acid/DNA or RNA enters the cell.	1
7	Virus or nucleic acid alters the cells metabolism.	1
8	Copies of the viral nucleic acid are made.	1
9	The viral protein coat is synthesised.	1
10	The virus uses the host cells nucleotides/amino acids or ATP.	1
11	The viruses are then assembled or the nucleic acids are surrounded by the protein coats.	1
12	Large numbers of viruses are released by lysis of the cell.	1
	Any to a maximum of 6 marks	**6**

Coherence	1	Writing must be under sub-headings e.g. Nature of Viruses and Replication or divided into paragraphs.	
	2	Related information must be grouped together with at least one point from nature of viruses and 4 points from viral replication.	
		Both points needed	**1**
Relevance	1	Must not give details of cellular defence mechanisms.	
	2	Must have at least one relevant point from nature of viruses and 4 points from viral replication.	
		Both points needed	**1**
			10

Answers continued

2.1 Variation

Answer to SBQ 1

(a) D, B, A, C, E — 1

(b) Chiasma(ta) — 1

(c) Independent assortment, crossing over (both) — 1

Answer to SBQ 2

M, K, J, L — 1

Top Tip

When working out allele positions from recombination frequencies, draw a working line to represent the chromosome – like this:

M K J L

Answer to SBQ 3

(a) Male BbCc, female bbCc — 1

Getting these answers right needs a bit of detective work – look at the clues in the question. The male has a black coat so must have the B allele but since he has produced offspring with red coats he must also have a b allele. The female must be bb since she has a red coat. Both parents had plain coats so each must have at least one C allele. They have produced spotted coat offspring so they must each have a c allele. Problem solved!

(b) 3 plain black : 3 plain red : 1 spotted black : 1 spotted red — 1

This answer is dependent on getting part (a) correct – in real exams, questions are not allowed to be dependent. Notice that the ratio is not typical of dihybrid crosses in general – it shows that you must always do the working in cross problems.

Answer to SBQ 4

(a) Y – X^DX^d, Z – X^dY (both) — 1

(b) zero % — 1

(c) Q has received one deficiency allele from her father.

Q has received the other deficiency allele from her mother.

Q's mother was a carrier. — 1

Answers continued

Top Tip

The question started with 'Explain' and there are actually three parts to the **full** answer – a sure sign of an A-type question!

Answer to SCQ 1

(i)	1	Random/spontaneous/ by chance.	1
	2	Low frequency or rare.	1
	3	Example of a chemical mutagenic agent e.g. mustard gas.	1
	4	Example of irradiation e.g. X-rays, UV light.	1
	5	Mutagenic agents increase rate/frequency/chance of mutation.	1
		Any to a maximum of 3 marks	**3**
(ii)	6	Gene mutation is a change in the base types/sequence/order.	1
	7	Substitution – one base replaced with another.	1
	8	Inversion – order of bases reversed.	1
	9	Substitution/inversion may change base order of one or two codons only.	1
	10	Substitution/inversion may change only one/two amino acid(s).	1
	11	Deletion – bases removed/taken out.	1
	12	Insertion – bases added/put in.	1
	13	Deletion/insertion changes all codons after the mutation.	1
	14	Deletion/insertion changes all amino acids after the mutation.	1
	15	Protein made after substitution/inversion may still work/function.	1
		OR protein made after deletion/insertion will not work/function.	1
		Any to a maximum of 7 marks	**7**
			10

Answer to SCQ 2

1	Number changes due to non-disjunction.	1
2	Failure of chromosomes to segregate/move apart normally.	1
3	Duplication – part of a chromosome repeated.	1
4	Translocation – part of one chromosome becomes joined to another.	1
5	Deletion – part of a chromosome missing/lost.	1
6	Inversion – part of a chromosome turned round/reversed.	1
	Any to a maximum of 4 marks	**4**

Answers continued

7	Polyploidy – individual possesses one or more sets of chromosomes above normal.	1
8	Normal number is diploid.	1
9	Polyploidy often confers increased vigour to plants.	1
10	One example of vigour: high yield/fast growth.	1
11	Another example of vigour: resistance to frost/drought/disease.	1
12	Has been induced artificially in crop plants.	1
	Any to a maximum of 4 marks	**4**

Coherence	1	Writing must be under sub-headings e.g. Changes in Chromosome Number and Structure and Polyploidy or divided into paragraphs.	
	2	Related information must be grouped together with at least two points from each section.	
		Both points needed	**1**
Relevance	1	Must not give details of gene mutation, hybridisation, somatic fusion etc.	
	2	Must have at least two relevant points from each section.	
		Both points needed	**1**
			10

2.2 Selection and Speciation

Answer to SBQ 1

(a)	Have differently shaped/sized beaks as they eat different food/feed differently	1
	Note that the actual difference between beaks is needed and this must be related to feeding method or food type eaten.	
(b)	Evidence of a common ancestor	1
(c)	Ecological, reproductive (both needed)	1

Top Tip

Remember GERM?

(d)	Could not interbreed to produce fertile young	1
(e)	Wildlife reserves, captive breeding, cell banks	
	Any 2	2

Answers continued

Answer to SBQ 2

(a)	Gene probes/use of banding patterns	1
(b) 1	endonuclease	1
2	ligase	1
(c)	Insulin/growth hormone	1

Answer to SCQ 1

(i)	1	A species is a group of organisms interbreeding to produce fertile young.	1
	2	Mention of a common gene pool.	1
	3	Species/population separated into two by isolating barrier/mechanism.	1
	4	Prevents gene flow/exchange.	1
	5	Two isolation mechanisms given e.g. geographical/ecological.	1
	6	A third type of isolation mechanism given e.g. reproductive.	1
		Any to a maximum of 4 marks	**4**
(ii)	7	Mutations in each sub-population different.	1
	8	Mutation gives new genes/alleles/variation/phenotypes/gene pool.	1
	9	The sub-population are in different environments.	1
	10	Natural selection pressures different for each sub-population.	1
	11	Best adapted/suited survive OR survival of fittest.	1
	12	They pass on favourable genes/alleles/characteristics to offspring.	1
	13	Over many generations/long period of time.	1
	14	New species formed when groups can no longer interbreed to produce fertile young.	1
		Any to a maximum of 6 marks	**6**
			10

Answer to SCQ 2

1	Selective breeding undertaken by humans.	1
2	Parents are chosen because of their desirable characteristics.	1
3	One example of a species with desirable characteristic e.g. cattle with high milk yield	1
4	Another example e.g. wheat with increased grain yield.	
5	Offspring show improvement of desired characteristic over long periods.	1
6	Hybridisation is the crossing of two different species to combine their desirable characteristics together.	1
	Any to a maximum of 4 marks	**4**
7	Somatic fusion overcomes sexual incompatibility between plant species.	1
8	Fusion of non-gamete cells e.g. leaf cells.	1

Answers continued

9	Cell wall of somatic cells removed by cellulose.	1
10	Fusion of protoplasts to form hybrid.	1
11	Hybrid cultured to form new plant.	1
12	New plant has characteristics of both parent varieties, or example of this.	1
	Any to a maximum of 4 marks	**4**

Coherence	1	Writing must be under sub-headings e.g. Selective Breeding and Hybridisation and Somatic Fusion (separate sub-heading for Hybridisation acceptable but not required) or divided into paragraphs.	
	2	Related information must be grouped together with at least two points from each section.	
		Both points needed	**1**
Relevance	1	Must not give details of natural selection, genetic engineering.	
	2	Must have at least two relevant points from each section.	
		Both points needed	**1**
			10

2.3 Animal and Plant Adaptations

Answer to SBQ 1

(a)	From high to low water concentration/from hypotonic to hypertonic	1
(b)	Has a larger/greater surface area for osmosis	1
(c)	Adhesion	1
(d)	Guard cells lose turgor/become flaccid.	1
(e)	Provides water for support or provides mineral/nutrients or cooling/heat loss	
	Any 1 = 1	
	Any 2	2

Answer to SBQ 2

(a)	(i)	1	Avoidance behaviour/escape response	1
		2	Defence against predators/protection/prevent damage	1
	(ii)	1	Habituation	1
		2	Saves energy or can continue feeding	1
(b)	Thorns, spines (both)			1

Answer to SCQ 1

(i)	1	Animals use energy in hunting/foraging for food.	1
	2	Energy gain from food must exceed energy used in hunting/foraging.	1

Answers continued

	3	Behaviour adopted gives maximum net gain of energy.	1
	4	Any two examples of foraging behaviour, from:	
		◆ a description of a search pattern/technique	
		◆ selection of prey item size	
		◆ response to chemical stimulation	2
		Any to a maximum of 4 marks	**4**
(ii)	5	In co-operative hunting a group hunts together.	1
	6	An example of the benefits, from:	
		◆ reduced energy expenditure by individual	
		◆ increased hunting success	
		◆ larger prey item can be hunted	
		◆ food sharing includes weaker/younger individuals	1
	7	Dominance hierarchy has a rank order or a description of rank order.	1
	8	An example of the benefits, from:	
		◆ dominant individuals are more experienced and increase hunting success	
		◆ subordinate individuals share food	1
	9	Territorial behaviour involves defence of a specific area.	1
	10	Occupants use territory exclusively for obtaining food.	1
		Any to a maximum of 4 marks	**4**
	11	Social group can attack/mob a predator, or example of this.	1
	12	Predators can be confused by/unable to select an individual in a large moving group.	1
	13	Members of a group can share in the watch for predators.	1
	14	Members of a group can protect young.	1
		Any to a maximum of 2 marks	**2**
			10

Answer to SCQ 2

1	Blood/body fluids at higher water concentration/hypotonic to surroundings.	1
2	Water lost by osmosis from gills/mouth.	1
	Either to maximum of 1	**1**
3	Drinks sea water to replace loss.	1
4	Salts removed by chloride secretory cells.	1
5	This process requires active transport/energy.	1
6	Few glomeruli in kidney.	1
7	Small glomeruli in kidney.	1

Answers continued

8 Low filtration rate. 1
9 High rate of water reabsorption. 1
10 Small volume of urine produced. 1
11 Dilute urine produced. 1
Any to a maximum of 7 **8**

Coherence
1 Writing must be under sub-headings e.g. Osmotic Problems and Water Balance or divided into paragraphs.
2 Related information must be grouped together with at least one point from 1–2 and four points from 3–11.
Both points needed **1**

Relevance
1 Must not give details of freshwater fish.
2 Must have at least one point from 1–2 and four points from 3–11.
Both points needed **1**

10

3.1 Control of Growth and Development

Answer to SBQ 1

(a) It produces new cells. 1

Do not be tempted to say that it produces new xylem. The cambium cells actually undergo cell division producing undifferentiated cells that can then differentiate to form new xylem or phloem.

(b) Spring xylem vessels are wider or have a larger diameter. 1

(c) Different or specific genes are active or switched on while other genes remain switched off. 1

*It is not acceptable to say than genes are switched on **and** off.*

Answer to SBQ 2

Insect moults or sheds its skin or exoskeleton allowing it to grow and increase in length. = 1 mark

Its skin or exoskeleton then hardens preventing any further growth. = 1 mark 2

Shell or coat is not acceptable as an alternative to skin or exoskeleton.

Answer to SBQ 3

(a) Genes control or code for the synthesis of the enzymes which carry out the chemical reactions in the metabolic pathway.

(b) Gene mutation results in a change in the base order. = 1 mark

This results in a change in the order of the amino acids in the protein being coded for so an abnormal enzyme is produced. = 1 mark 2

Answers continued

Top Tip

Learn this content as:

- order of bases determines
- order of amino acids that determines
- specific enzyme in a metabolic pathway

Answer to SBQ 4

(a) Repressor molecule or repressor protein binds to and blocks the operator which prevents it switching on the structural gene. 1

(b) Lactose combines with the repressor molecule. This means that the operator is free to switch on the structural gene. 1

Top Tip

Remember ROSE?

(c) Saves energy or it only makes the enzyme when it is needed. 1

Answer to SBQ 5

(a)

Process	IAA	GA	Both
Stimulates amylase production barley grains		✓	
Inhibits leaf abscission	✓		
Promotes fruit formation	✓		
Stimulates stem elongation			✓
Breaks bud dormancy		✓	
Causes apical dominance	✓		

Two or three correct = 1 mark

Four or five correct = 2 marks

All six correct 3

Top Tip

Remember MED-C, OLAF and GA-AID?

Answers continued

(b) Herbicide or selective weedkiller; rooting powder; production of seedless fruits 1

(c) To make amino acids, proteins, enzymes, DNA, RNA, or chlorophyll 1

Answer to SCQ 1

1	Nitrogen is required for protein or amino acid synthesis.	1
2	Protein is required for enzymes or membranes.	1
3	Nitrogen is required for DNA or RNA.	1
4	DNA essential for mitosis and cell division or DNA and RNA essential for protein synthesis.	1
5	Nitrogen is required for chlorophyll.	1
6*	Chlorophyll is needed for photosynthesis.	1
7	Symptoms of nitrogen deficiency include: reduced shoot growth, chlorosis, red leaf bases and longer root systems.	
	Any 3 correct = 2 marks; any 2 correct = 1 mark	2
	Any to a maximum of 5 marks	**5**
8	Magnesium is required for chlorophyll.	1
9*	Chlorophyll is needed for photosynthesis.	1
10	Plants lacking magnesium shows overall reduced growth and chlorosis.	1
	A maximum of 1 mark is available for chlorophyll deficiency	**1**
11	Vitamin D is required for the uptake of calcium from the intestines into the blood.	1
12	Vitamin D is required for the uptake of calcium into the bones.	1
13	Deficiency in vitamin D leads to rickets.	1
	A maximum of 2 marks can be gained for Vitamin D	**2**
14	Iron is a component of haemoglobin, many enzymes or hydrogen-carrying systems e.g. cytochromes.	2
15	Deficiency in iron leads to anaemia.	1
	A maximum of 2 marks can be gained for iron	**2**
		10

*award only one

Answer to SCQ 2

1	Plants or shoots show phototropism.	1
2	Light coming from one side results in a greater concentration of IAA on the dark side.	1
3	This results in greater elongation of cells on the dark side.	1
4	Etiolation results in the absence of light.	1

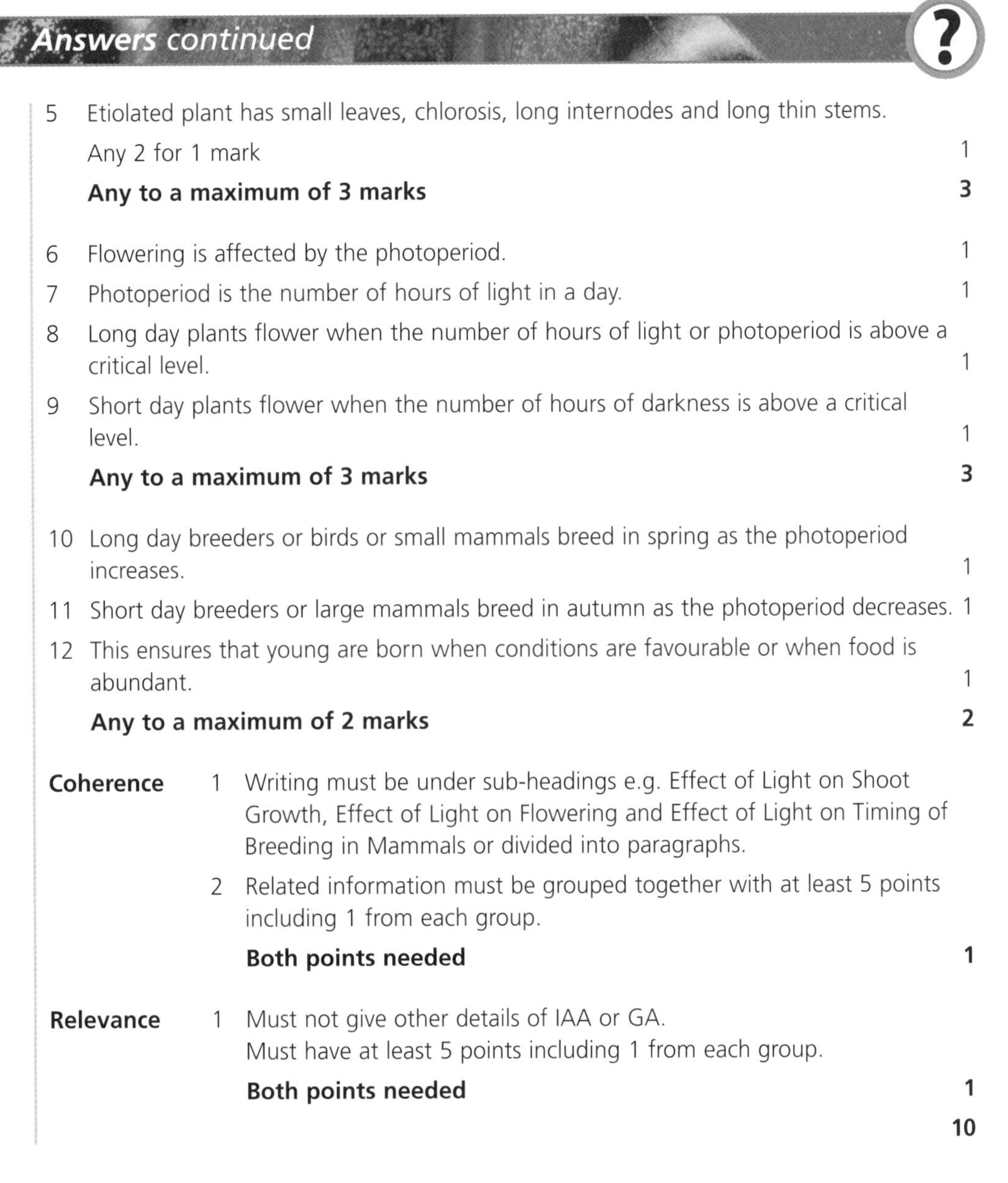

Answers continued

		Marks
5	Etiolated plant has small leaves, chlorosis, long internodes and long thin stems.	
	Any 2 for 1 mark	1
	Any to a maximum of 3 marks	**3**
6	Flowering is affected by the photoperiod.	1
7	Photoperiod is the number of hours of light in a day.	1
8	Long day plants flower when the number of hours of light or photoperiod is above a critical level.	1
9	Short day plants flower when the number of hours of darkness is above a critical level.	1
	Any to a maximum of 3 marks	**3**
10	Long day breeders or birds or small mammals breed in spring as the photoperiod increases.	1
11	Short day breeders or large mammals breed in autumn as the photoperiod decreases.	1
12	This ensures that young are born when conditions are favourable or when food is abundant.	1
	Any to a maximum of 2 marks	**2**

			Marks
Coherence	1	Writing must be under sub-headings e.g. Effect of Light on Shoot Growth, Effect of Light on Flowering and Effect of Light on Timing of Breeding in Mammals or divided into paragraphs.	
	2	Related information must be grouped together with at least 5 points including 1 from each group.	
		Both points needed	**1**
Relevance	1	Must not give other details of IAA or GA. Must have at least 5 points including 1 from each group.	
		Both points needed	**1**
			10

3.2 Physiological Homeostasis

Answer to SBQ 1

(a)	Hypothalamus	1
(b)	Vasoconstriction; decreased (both)	1
(c)	Through nerves or neurones or nervous system	1
(d)	Allows enzymes to react or work at their optimum temperature	1

Answers continued

Answer to SBQ 2

(a) The corrective mechanisms are switched on due to a change being detected. =1
When the correction is made the corrective mechanism is switched off. =1 2
This works as an A- type question because candidates often get the first mark only.

(b) The pancreas 1

(c) Hormone X – insulin; hormone Y – glucagon (both) 1

(d) Organ – liver; storage carbohydrate – glycogen (1 mark for each correct) 2

Top Tip

Remember the rhyme?

Low blood sugar – glucose gone
What you need is glucagon
To turn glucose into glycogen
What you need is insulin

(e) Blood glucose concentration increases. 1
Adrenaline converts the stored glycogen into glucose.

Answer to SCQ 1

(i)
1 The water concentration is detected by the hypothalamus. 1
2 A decrease in the water concentration leads to an increase in the production of ADH. 1
3 ADH is transported in the blood to the kidneys. 1
4 ADH increases the permeability of the tubules. 1
5 This results in increased water reabsorption. 1
6 A low volume of concentrated urine is produced. 1
7 The water concentration is returned to normal. 1
8 The ADH production is then decreased. 1
Any to a maximum of 5 marks **5**

(ii)
9 Blood glucose concentration is detected by the pancreas. 1
10 If the blood glucose concentration increases the secretion of insulin increases. 1
11 Insulin increases the permeability of the liver cells to glucose or liver cells take up glucose. 1
12 Glucose is converted to glycogen. 1
13 If the blood glucose concentration decreases the secretion of glucagon increases. 1
14 Glucagon causes glycogen to be converted to glucose. 1

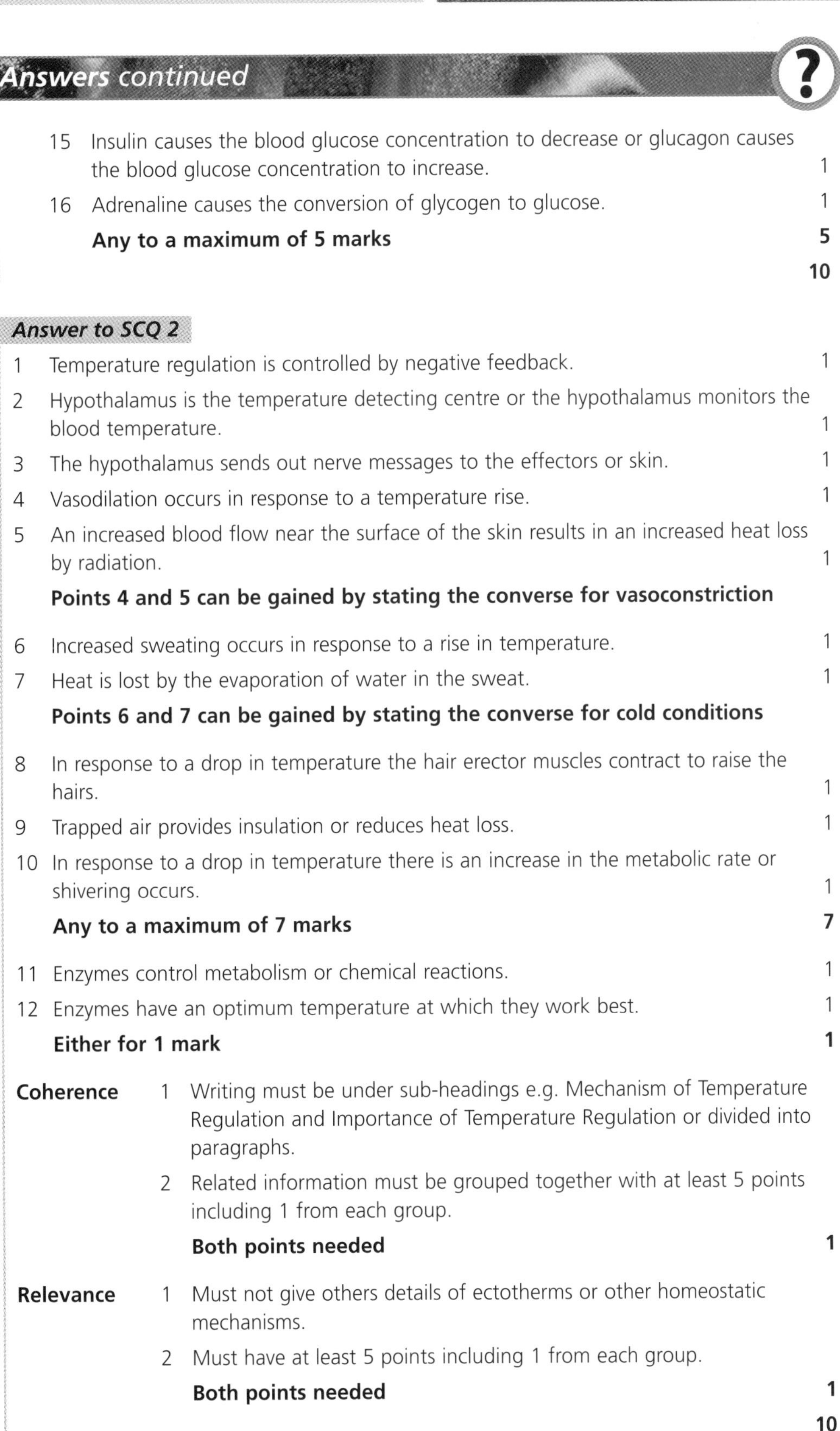

Answers continued

15 Insulin causes the blood glucose concentration to decrease or glucagon causes the blood glucose concentration to increase. 1

16 Adrenaline causes the conversion of glycogen to glucose. 1

Any to a maximum of 5 marks **5**

10

Answer to SCQ 2

1 Temperature regulation is controlled by negative feedback. 1

2 Hypothalamus is the temperature detecting centre or the hypothalamus monitors the blood temperature. 1

3 The hypothalamus sends out nerve messages to the effectors or skin. 1

4 Vasodilation occurs in response to a temperature rise. 1

5 An increased blood flow near the surface of the skin results in an increased heat loss by radiation. 1

Points 4 and 5 can be gained by stating the converse for vasoconstriction

6 Increased sweating occurs in response to a rise in temperature. 1

7 Heat is lost by the evaporation of water in the sweat. 1

Points 6 and 7 can be gained by stating the converse for cold conditions

8 In response to a drop in temperature the hair erector muscles contract to raise the hairs. 1

9 Trapped air provides insulation or reduces heat loss. 1

10 In response to a drop in temperature there is an increase in the metabolic rate or shivering occurs. 1

Any to a maximum of 7 marks **7**

11 Enzymes control metabolism or chemical reactions. 1

12 Enzymes have an optimum temperature at which they work best. 1

Either for 1 mark **1**

Coherence

1 Writing must be under sub-headings e.g. Mechanism of Temperature Regulation and Importance of Temperature Regulation or divided into paragraphs.

2 Related information must be grouped together with at least 5 points including 1 from each group.

Both points needed **1**

Relevance

1 Must not give others details of ectotherms or other homeostatic mechanisms.

2 Must have at least 5 points including 1 from each group.

Both points needed **1**

10

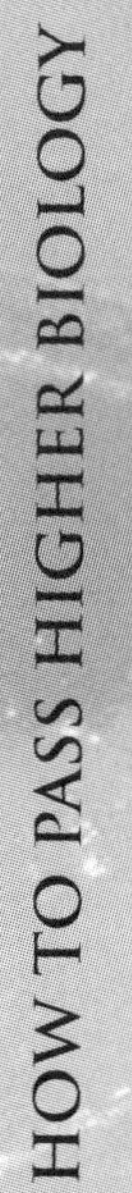

3.3 Population Dynamics

Answer to SBQ 1

(a) Biological indicators of pollution / conservation of endangered species/control of pests
Any 2 — 1

(b) Predation/food supply/disease or competition for space
Any 2 — 1

Answer to SBQ 2

(a) Succession — 1

(b) Increased biomass/increased species diversity/increased stability/more complex food webs
Any 2 — 2

(c) Presence changes the soil by increasing or improving the fertility or mineral content or humus or organic material/the soil depth/water retention/drainage/pH/aeration/stability.
Any 2 — 1

Answer to SCQ 1

(i)
1. The management of food and raw materials — 1
2. The control of pest species — 1
3. The assessment of pollution levels — 1
4. The protection and conservation of endangered species — 1
5. Also provides data to measure the effectiveness of any control measures adopted — 1

Any to a maximum of 4 marks — **4**

(ii)
6. Factors affecting the size of a population fall into two categories called density independent and density dependent factors. — 1
7. Density independent factors are factors which affect the population regardless of the population density. — 1
8. Examples of density independent factors include extremes in temperature and rainfall such as high or freezing temperatures, or floods or drought.
Any 2 — 2
9. Effect of density dependent factors increases as the population density increases. — 1
10. Examples of density dependent factors include disease, food supply, predation and competition for space.
Any 2 — 2
11. Relate factors to the effect on the population:

Answers continued

- When food is in short supply, competition occurs between individuals and some may starve to death.
- If the population increases the predation also increases. Predators are likely to kill more of a dense population than a widely-spread one.
- If the population increases then there is an increase in the spread of disease.
- An increase in population results in increased competition for space or territory.

Any 2 | 2

Any to a maximum of 6 marks | **6**

10

Answer to SCQ 2

1 Plant succession is the gradual but regular change in the species of plants present in a particular habitat. 1

2 Succession is unidirectional or succession causes habitat modification. 1

3 First species to move into an area are called primary colonisers or pioneer species, or named examples of pioneer species e.g. lichen, mosses, herbs, grasses and weeds. 1

4 Early colonisers are wind dispersed or have a low nutrient requirements or have a shallow root system or are adapted to adverse conditions. 1

5 Their presence changes the soil by increasing the soil fertility and the depth of the soil, and they improve the water retention, pH, drainage and aeration. Any 1 1

6 Other species move in and repeated take-overs/colonising occurs until the climax (final) community is reached. 1

7 Characteristics of the climax community are that it has a greater biomass, a greater species diversity, a greater stability or more complex food webs. Any 2 1

Any to a maximum of 4 marks **4**

8 Effect of density dependent factors increases as the population density increases. 1

9 Examples of density dependent factors include disease, food supply, predation and competition for space. Any 2 for 1 mark 1

10 When food is short supply, competition occurs between individuals and some may starve to death.

If the population increases the predation also increases. Predators are likely to kill more of a dense population than a widely-spread one.

If the population increases then there is an increase in the spread of disease.

An increase in population results in increased competition for space or territory.

Any 2 2

11 Such changes will cause the population to decrease. 1

12 Population returns to the size that the environment can support. 1

Any to a maximum of 4 marks **4**

Answers continued

Coherence	1	Writing must be under sub-headings e.g. Succession and Climax and Density Dependant Factors or divided into paragraphs.	
	2	Related information must be grouped together with at least 5 points including 2 from each group.	
		Both points needed	**1**
Relevance	1	Must not give other details of density independent factors.	
	2	Must have at least 5 points including 2 from each group.	
		Both points needed	**1**
			10

Chapter 8

PROBLEM SOLVING: ANSWERS AND COMMENTARY

Answers

Answer to 1 Data Handling

(i) (a) 104 beats/minute

146 cm^3 — 1

Hopefully you remembered to use a ruler on the graph to draw a line up from 1·4 on the x-axis on to the line for the pulse rate and over to the left-hand y-axis and then up to the line for the stroke volume and then to the right to obtain the correct answers.

(b) The pulse rate increases from 110 to 170 beats/minute. =1 The stroke volume rises from 150 to 156 cm^3 at 2·0 litres/minute and then remains constant. =1 — 2

The best advice for this type of question is to make sure that you describe the changes fully, giving the figures and units from each axis.

(c) 8·14 litres — 1

From the graph you should see that a rate of oxygen uptake of 0·4 litres/minute corresponds to a stroke volume of 110 cm^3 (volume of blood pumped out of the heart in one beat) and a pulse rate of 74 beats per minute. From this you can calculate the volume of blood leaving the heart in one minute as 74×110 cm^3 = 8140 cm^3. This then has to be divided by 1000 to convert it to litres, giving the answer as 8·14 litres.

(d) 1·1 litres oxygen/minute — 1

$$80 = 0{\cdot}5 \text{ and } 130 = 1{\cdot}6$$

By subtracting, this gives an increase of 1·1.

(e) 25% — 1

$$\text{Percentage increase} = \frac{\text{increase}}{\text{starting value}} \times 100$$

$$= \frac{0{\cdot}4}{1{\cdot}6} \times 100 = 25$$

(f) 2 : 3 — 1

The values from the graph for the pulse rate and stroke volume when the oxygen uptake is 0·6 litres/minute are 80 and 120 respectively.

Answers continued

These must then be simplified as describe earlier. Dividing by 10 takes them to 8 and 12. However, they can be simplified further by dividing by 4 to give our answer of 2 : 3.

(ii) (a) 6 litres — 1

From the graph we can see that when the rate of oxygen uptake is 1·6 litres/minute, there are 30 litres of air inhaled.

$$20\% \text{ of } 30 = \frac{20}{100} \times 30 \qquad \text{or } 30 \times 20 \text{ (press \% button).}$$

(b) 30 litres inhaled/minute — 1

This type of question requires you to use both graphs to obtain the answer.

By checking the space for your answer on the question paper you would see from the units given which axis would give you the final answer.

Another tip is that there will always be a label on each graph which is the same. This label connects them and allows you to move from one graph to the next. The general pattern is a movement from the x-axis up and then left over to the y-axis to obtain a value and then move from the second graph from the y-axis right and then down on to the x-axis to get the answer. In this example we start at 110 on the y-axis on Graph 1, move right, then down onto the x-axis to 1·6 litres/minute. Next we take this value and start on the x-axis of the second graph and move up and then left onto the y-axis to get our answer of 30 litres inhaled/minute.

Answer to 2 Experimental Procedure

(a) For the first mark the following criteria must be met.

Size — the graph must fill more than half of your piece of paper.

Scale — must allow points to be plotted accurately.

Labels — must be copied exactly from the table.

Units — must be included and in this example the negative values must be shown. — 1

For the second mark the points must be plotted correctly and connected using a ruler. Any sloppy work will be penalised so this should be done carefully. — 1

(b) *Only the variable being investigated should be altered. All other variables should be controlled.*
In this example, the concentration was being altered. The other variables, which should therefore be kept constant, could relate to the solution or the tissue.

Solution: the volume, temperature, pH and the same type of salt. — 1

Tissue: the same potato, surface area, shape. — 1

Answers continued

(c) The reason the tissues are immersed for one hour is:

to allow time for any change in mass to occur

to allow time for a measurable change

to allow time for osmosis/diffusion to occur.

Any one 1

Questions regarding time are fairly common and it is good practice to refer to named chemicals or actual processes in your answer if possible.

(d) Blotting is carried out to remove surface water.
This means that any recorded change in mass is due to a change in the cell contents or due to the movement of water into or out of the tissue. 1

(e) The starting masses were different.
It allows a fair comparison to be made when the starting masses are different. 1

Results should be converted into a percentage whenever the starting measurements are different.

(f) Repeat the experiment or use more tissues at each concentration. 1

Top Tip

Remember ***ROAR*** = ***R****epeat* ***O****btain an* ***A****verage to increase* ***R****eliability*

(g) As the salt solution concentration increased, the mass of water lost by osmosis increased. 1

The answer relates the trend in the results to the process of osmosis under investigation as mentioned in the question stem.

(h) –20% 1

This is obtained by continuing the graph until it intersects the 0·4 M line. It is OK to add the extra part to the line on the graph you have drawn.